Phase Transformations and Material Instabilities in Solids

Publication No. 52
of the Mathematics Research Center
The University of Wisconsin—Madison

Academic Press Rapid Manuscript Reproduction

Phase Transformations and Material Instabilities in Solids

Edited by

Morton E. Gurtin

Department of Mathematics
Carnegie—Mellon University
Pittsburgh, Pennsylvania

Proceedings of a Conference
Conducted by the Mathematics Research Center
The University of Wisconsin—Madison
October 11–13, 1983

1984

Academic Press
(Harcourt Brace Jovanovich, Publishers)
Orlando San Diego New York London
Toronto Montreal Sydney Tokyo

ACADEMIC PRESS, INC.
Orlando, Florida 32887

United Kingdom Edition published by
ACADEMIC PRESS, INC. (LONDON) LTD.
24/28 Oval Road, London NW1 7DX

Library of Congress Cataloging in Publication Data

Main entry under title:

Phase transformations and material instabilities in solids.

Includes index.
1. Phase transformations (Statistical physics)--Congresses. 2. Solids--Surfaces--Congresses. I. Gurtin, Morton E. II. University of Wisconsin--Madison. Mathematics Research Center. III. Title: Material instabilities in solids.
QC176.8.P45P43 1984 530.4'1 84-45860
ISBN 0-12-309770-3 (alk. paper)

PRINTED IN THE UNITED STATES OF AMERICA

84 85 86 87 9 8 7 6 5 4 3 2 1

Contents

CONTRIBUTORS

Numbers in parentheses indicate the pages on which the authors' contributions begin.

J. M. BALL (1), *Department of Mathematics, Heriot-Watt University, Edinburgh, Scotland*

E. CHATER (21), *Division of Applied Sciences, Harvard University, Cambridge, Massachusetts 02138*

D. A. DREW (37), *Department of Mathematical Sciences, Rensselaer Polytechnic Institute, Troy, New York 12181*

J. L. ERICKSEN (61), *Department of Aerospace Engineering and Mechanics, School of Mathematics, University of Minnesota, Minneapolis, Minnesota 55455*

J. E. FLAHERTY (37), *Department of Mathematical Sciences, Rensselaer Polytechnic Institute, Troy, New York 12181*

M. E. GURTIN (99), *Department of Mathematics, Carnegie-Mellon University, Pittsburgh, Pennsylvania 15213*

R. HAGAN (113), *Department of Mathematics, University of Oregon, Eugene, Oregon 97403*

J. W. HUTCHINSON (21), *Division of Applied Sciences, Harvard University, Cambridge, Massachusetts 02138*

R. D. JAMES (79), *Division of Engineering, Brown University, Providence, Rhode Island 02912*

J. S. LANGER (129), *Institute for Theoretical Physics, University of California, Santa Barbara, Santa Barbara, California 93106*

R. F. SEKERKA (147), *Mellon College of Science, Carnegie-Mellon University, Pittsburgh, Pennsylvania 15213*

J. SERRIN (113), *School of Mathematics, University of Minnesota, Minneapolis, Minnesota 55455*

J. E. TAYLOR (205), *Mathematics Department, Rutgers University, New Brunswick, New Jersey 08903*

M. SLEMROD (163), *Department of Mathematical Sciences, Rensselaer Polytechnic Institute, Troy, New York 12181*

PREFACE

An interdisciplinary conference on phase transitions and material instabilities in solids was held at the Mathematics Research Center, University of Wisconsin–Madison in October, 1983. The conference was sponsored by the National Science Foundation under Grant No. MCS-8210950 and by the United States Army under Contract No. DAAG29-80-C-0041. This volume collects the invited talks, whose topics include general theories of phase transitions, equilibrium shapes of surfaces, morphological instabilities and dendrite formation, shock-induced phase transitions, and related results on the calculus of variations.

I would like to thank Professor John Nohel for his encouragement, support, and assistance in planning this conference; the National Science Foundation and the Army Research Office for their financial support; Mrs. Judith Siesen for her assistance in editing this volume; and Mrs. Gladys Moran for her help in organizing the conference.

MORTON E. GURTIN

MATERIAL INSTABILITIES AND THE CALCULUS OF VARIATIONS

J. M. Ball

1. INTRODUCTION.

The aim of the calculus of variations is to study the minimization of integrals depending on unknown functions. In continuum mechanics a common procedure is to minimize a 'free energy' integral, the minimizing functions being interpreted as equilibrium displacement and temperature fields. The motivation lies in thermodynamics. Roughly, we seek an appropriate Lyapunov function for the governing equations, typically of the form

$$E(u) = \int_{\text{body}} \mathscr{F}(X, J^k u(X,t))\,dX,$$

where u is a vector of field variables (displacement, velocity, density, temperature etc.) and $J^k u$ denotes the set of all partial derivatives of u with respect to X of all orders r with $0 \leqslant r \leqslant k$; that is, $E(u(\cdot,t))$ is a non-increasing function of time t along solutions. Often we add the extra requirement that $E(u(\cdot,t))$ is constant if and only if $u = u(X)$ is a time-independent solution. In general there may be many time-independent solutions, infinitely many in the case of some problems involving phase transitions, leading to complicated behaviour of solutions as $t \longrightarrow \infty$. Some solutions may have atypical asymptotic behaviour, converging, for example, to unstable time-independent solutions. However, in the presence of a Lyapunov function E we expect that such exceptional solutions will lie in a negligible

ISBN 0-12-309770-3

subset N of the phase space X of admissible functions. We further expect that the remainder $X\setminus N$ of the phase space is the disjoint union of 'larger' positively invariant sets S_α and that solution paths u in S_α are minimizing for E, i.e.

$$\lim_{t\to\infty} E(u(\cdot,t)) = \inf_{v\in S_\alpha} E(v).$$

In particular, if $t_j \longrightarrow \infty$ then $v_j(x) = u(x,t_j)$ will be a <u>minimizing sequence</u> for E in S_α, i.e.

$$E(v_j) \searrow \inf_{v\in S_\alpha} E(v).$$

In especially favourable cases there may be just one $S_\alpha = X$ with N empty and all solution paths minimizing for E in X. In general a particular S_α might contain a number of time-independent solutions with the same value of E, or no time-independent solution at all.

For specific problems the following natural questions are important:

(Q1) Do the governing equations admit one or more nontrivial Lyapunov function E ?

(Q2) Given an appropriate subset S of X, does E attain a minimum on S ?

(Q3) What conditions does a minimizer satisfy?

(Q4) Do all minimizing sequences for E on S tend to minimizers? If not, what happens?

(Q5) When <u>is</u> $u(\cdot,t_j)$ a minimizing sequence, and what special properties do such sequences, realized by the dynamics, possess?

(Q6) What can be said about the structure of the decomposition $X = N \cup \bigcup_\alpha S_\alpha$?

These questions are particularly interesting for materials which can undergo phase transitions; typically the governing equations can then change type (cf Ericksen [20]). In this article we make some remarks concerning the first four questions but say nothing about the last two, about which little is known. (Some partial, but inconclusive, information about (Q5) was obtained in a model problem by Andrews & Ball [2].)

2. LYAPUNOV FUNCTIONS IN NONLINEAR THERMOELASTICITY.

We address (Q1) - (Q3) in the context of a nonlinear thermoelastic material. The results are taken from joint work with G. Knowles [6] that is still in progress. Some of the calculations are formal, and no attempt is made to make precise all the hypotheses concerning regularity etc.

We are concerned with a thermoelastic material occupying the bounded strongly Lipschitz open subset $\Omega \subset \mathbb{R}^n$ in a reference configuration. At time t the particle occupying the point $X \in \Omega$ in the reference configuration has position $x(X,t) \in \mathbb{R}^n$ and temperature $\theta(X,t) > 0$. For simplicity we suppose that there is no external body force or heat supply. The governing equations are then

$$\rho_R \ddot{x} = \operatorname{Div} T_R, \tag{2.1}$$

$$\rho_R \dot{U} - \operatorname{tr}(T_R \dot{F}^T) + \operatorname{Div} q_R = 0, \tag{2.2}$$

where $\rho_R(X)$ is the density in the reference configuration, T_R is the Piola-Kirchhoff stress tensor, U is the internal energy density, $F = \nabla x(X,t)$ is the deformation gradient, and q_R is the (reference) heat flux vector. The constitutive relations are given in terms of the Helmholtz free energy $A(X,F,\theta)$ and specific entropy $\eta(X,F,\theta)$ by

$$\left.\begin{aligned} &T_R = \rho_R \frac{\partial A}{\partial F}, \quad \eta = -\frac{\partial A}{\partial \theta}, \quad U = A + \eta\theta, \\ &q_R = q_R(X,F,\theta,\operatorname{Grad}\theta). \end{aligned}\right\} \tag{2.3}$$

The second law of thermodynamics requires that

$$q_R \cdot \operatorname{Grad}\theta \leqslant 0, \tag{2.4}$$

and we shall assume that this inequality is strict for $\operatorname{Grad}\theta \neq 0$.

We impose the following boundary conditions:

$$\text{\underline{Mechanical}} \qquad \left.\begin{aligned} &x = x_0(X) \quad \text{on} \quad \partial\Omega_1, \\ &T_R N = 0 \quad \text{on} \quad \partial\Omega \setminus \partial\Omega_1. \end{aligned}\right\} \tag{2.5}$$

$$\text{\underline{Thermal}} \qquad \left.\begin{aligned} &\theta = \theta_0(X) \quad \text{on} \quad \partial\Omega_2, \\ &q_R \cdot N = 0 \quad \text{on} \quad \partial\Omega \setminus \partial\Omega_2. \end{aligned}\right\} \tag{2.6}$$

Here $\partial\Omega_1$, $\partial\Omega_2$ are given subsets of the boundary $\partial\Omega$, $N = N(X)$ is the unit outward normal to $\partial\Omega$ at X, and $x_o, \theta_o > 0$ are given functions.

We define $\mathscr{F} = \mathscr{F}(X,\dot{x},F,\theta)$ by

$$\mathscr{F} = \rho_R\left[\frac{1}{2}|\dot{x}|^2 + U - \phi(X)\eta\right],$$

where $\phi(X)$ is specified later.

A standard computation using (2.1) - (2.3) and (2.5) yields

$$\frac{d}{dt}\int_\Omega \mathscr{F}dX = \int_\Omega \left(\frac{\phi}{\theta} - 1\right)\operatorname{Div} q_R\, dX$$

$$= \int_{\partial\Omega}\left(\frac{\phi}{\theta} - 1\right) q_R\cdot N dA - \int_\Omega \left(\frac{\phi}{\theta}\right)_{,\alpha} q_R^\alpha dX. \qquad (2.7)$$

Provided that $\phi\big|_{\partial\Omega_2} = \theta_o\big|_{\partial\Omega_2}$ the surface integral vanishes by (2.6).

Special cases

1. Suppose θ_o is independent of X. In this case we choose $\phi \equiv \theta_o$ and (2.7) becomes, using (2.4),

$$\frac{d}{dt}\int_\Omega \mathscr{F}dX = \theta_o \int_\Omega \frac{q_R\cdot \operatorname{Grad}\theta}{\theta^2} dX \leqslant 0.$$

The result is well known; cf Duhem [16], Ericksen [18], Coleman & Dill [11], for example. The function $\mathscr{F} = \rho_R\left[\frac{1}{2}|\dot{x}|^2 + U - \theta_o\eta\right]$ is known as the equilibrium free energy.

2. Suppose that $q_R = q_R(X,\theta,\operatorname{Grad}\theta)$, and let ϕ satisfy the stationary heat equation

$$\operatorname{Div} q_R(X,\phi,\operatorname{Grad}\phi) = 0 \quad \text{in } \Omega \qquad (2.8)$$

with the same boundary conditions as θ, namely

$$\left.\begin{aligned} \phi &= \theta_o(X) && \text{on } \partial\Omega_2, \\ q_R(X,\phi,\operatorname{Grad}\phi)\cdot N &= 0 && \text{on } \partial\Omega\setminus\partial\Omega_2. \end{aligned}\right\} \qquad (2.9)$$

(In the examples considered below ϕ is unique.) By (2.7),

$\int_\Omega \mathcal{F} dX$ is a Lyapunov function provided

$$I(\theta) \overset{\text{def}}{=} \int_\Omega \left(\frac{\phi}{\theta}\right)_{,\alpha} q_R^\alpha(X,\theta,\text{Grad}\,\theta)\,dX \geqslant 0$$

for all $\theta > 0$ satisfying (2.6). It is easily verified that $\theta = \phi$ is a solution of the Euler-Lagrange equations for I. Since $I(\phi) = 0$ we are faced with a classical question in the calculus of variations, to decide if the given solution ϕ is a global minimum of I. The problem is not trivial because ϕ is only known implicitly and because the integrand may be negative. One interesting case which can be handled is when $q_R = -k(\theta)\text{Grad}\,\theta$, with the thermal conductivity $k(\theta)$ assumed positive. In this case $I(\theta) \geqslant 0$ if $\frac{\theta k'(\theta)}{k(\theta)}$ is a nonincreasing function of θ; conversely, if $\frac{\theta k'(\theta)}{k(\theta)}$ is non-decreasing and not constant then there exist domains Ω and boundary conditions (2.6) for which I may be negative. For the proofs and further results see [6]. To illustrate one of the methods for analyzing I consider the anisotropic linear case

$$q_R = -K(X)\text{Grad}\,\theta\,,$$

where the matrix $K(X)$ is positive for each X. Then letting $w = \log\theta - \log\phi$ we obtain

$$I(\theta) = \int_\Omega -\left(\frac{\phi}{\theta}\right)_{,\alpha} K^{\alpha\beta}\theta_{,\beta}dX$$

$$= \int_\Omega \left[\phi K^{\alpha\beta}w_{,\alpha}w_{,\beta} + K^{\alpha\beta}\phi_{,\beta}w_{,\alpha}\right]dX$$

$$\geqslant \int_\Omega K^{\alpha\beta}\phi_{,\beta}w_{,\alpha}dX = \int_\Omega \left[(K^{\alpha\beta}\phi_{,\beta}w)_{,\alpha} - w(K^{\alpha\beta}\phi_{,\beta})_{,\alpha}\right]dX$$

$$= \int_{\partial\Omega} K^{\alpha\beta}\phi_{,\beta}wN_\alpha dA = 0.$$

In particular, setting $T_R = 0$, $U = \theta$ we see that

$$\frac{d}{dt}\int_\Omega \rho_R(\theta - \phi\log\theta)\,dX \leqslant 0 \qquad (2.10)$$

for positive solutions θ, satisfying (2.6), of the linear heat equation

$$\rho_R \frac{\partial\theta}{\partial t} = \text{Div}(K(X)\text{Grad}\,\theta). \tag{2.11}$$

If $(x(\cdot), v(\cdot), \theta(\cdot))$ is a local minimum of

$$E(x,v,\theta) \overset{\text{def}}{=} \int_\Omega \rho_R\left[\frac{1}{2}|v|^2 + U(X,\nabla x,\theta) - \phi(X)\eta(X,\nabla x,\theta)\right]dX$$

subject to the boundary conditions (2.5),(2.6) then formally we have that

$$v = 0, \tag{2.12}$$

$$\rho_R\left(\frac{\partial U}{\partial\theta} - \phi\,\frac{\partial\eta}{\partial\theta}\right) = 0, \tag{2.13}$$

and

$$\text{Div}\;\rho_R\left(\frac{\partial U}{\partial F} - \phi\,\frac{\partial\eta}{\partial F}\right) = 0. \tag{2.14}$$

Using the thermodynamic identities (2.3) we obtain from (2.13) that

$$(\theta - \phi)\frac{\partial\eta}{\partial\theta} = 0,$$

which, assuming that the specific heat $\frac{\partial\eta}{\partial\theta}$ is positive, yields

$$\theta = \phi. \tag{2.15}$$

(This is what motivates the choice of ϕ in the special cases above.)

From (2.3),(2.14) and (2.15) we obtain

$$\text{Div}\;T_R = 0, \tag{2.16}$$

the usual equilibrium equation.

Special care has to be taken in the case when $\partial\Omega_2$ is empty, since then

$$\frac{d}{dt}\int_\Omega \rho_R\left(\frac{1}{2}|\dot{x}|^2 + U\right)dX = 0$$

for solutions of (2.1),(2.2),(2.5),(2.6), so that setting $v = \dot{x}$ we have

$$\int_\Omega \rho_R\left(\frac{1}{2}|v|^2 + U\right)dX = E_0, \tag{2.17}$$

where E_0 is a constant given by the initial data. Taking $\phi = 1$, it follows that $-\int_\Omega \rho_R\eta\,dX$ is a Lyapunov function. A local minimum of $-\int_\Omega \rho_R\eta\,dX$ subject to (2.17) and the boundary conditions (2.5),(2.6) formally satisfies

$$\rho_R\left(\frac{\partial\eta}{\partial\theta} - \lambda\,\frac{\partial U}{\partial\theta}\right) = 0\ , \tag{2.18}$$

$$\lambda\rho_R v = 0\ , \tag{2.19}$$

and

$$\mathrm{Div}\ \rho_R\left(\frac{\partial\eta}{\partial F} - \lambda\,\frac{\partial U}{\partial F}\right) = 0\ , \tag{2.20}$$

where λ is a Lagrange multiplier. If $\frac{\partial\eta}{\partial\theta} > 0$ then (cf Ericksen [18]) we deduce from (2.18) that

$$\theta = \frac{1}{\lambda} = \text{constant}\ ,$$

and thus $v = 0$ and (2.16) again holds. Similar considerations apply whenever the governing equations of a system possess conserved quantities (e.g. the mass constraint (3.8) below), and reinforce the need for a complete knowledge of all such conserved quantities.

Given appropriate existence theorems for minimizers (see [3,9]) it is not altogether obvious how to establish rigorously necessary conditions such as (2.16); some information on this question is given in [5].

3. MINIMIZERS AND MINIMIZING SEQUENCES FOR INVISCID FLUIDS WITH HEAT CONDUCTION.

In this section we consider (Q1) - (Q3), and especially (Q4), for an inviscid fluid with heat conduction. The results are taken from joint work with G. Knowles [6] that is still in progress and to which the reader is referred for a more detailed description. The fluid is assumed to be homogeneous and to occupy the spatial region $\omega \subset \mathbb{R}^n$, where ω is bounded and open. At time t and position $x \in \omega$ the fluid has density $\rho(x,t) \geqslant 0$, velocity $v(x,t) \in \mathbb{R}^n$ and temperature $\theta(x,t) > 0$. For simplicity we assume that there is no external body force or heat supply. The governing equations are then

$$\rho\dot{v} = -\ \mathrm{grad}\, p\ , \tag{3.1}$$

$$\dot{\rho} + \rho\,\mathrm{div}\, v = 0\ , \tag{3.2}$$

$$\rho\dot{U} + p\,\mathrm{div}\, v + \mathrm{div}\, q = 0\ , \tag{3.3}$$

where dots denote material time derivatives, p is the

pressure, U is the internal energy density and q is the (spatial) heat flux vector. The constitutive relations are given in terms of the Helmholtz free energy $A(\rho,\theta)$ and specific entropy $\eta(\rho,\theta)$ by

$$\left.\begin{aligned} &p = \rho^2 \frac{\partial A}{\partial \rho}, \quad \eta = -\frac{\partial A}{\partial \theta}, \quad U = A + \eta\theta, \\ &q = q(\rho,\theta,\operatorname{grad}\theta). \end{aligned}\right\} \tag{3.4}$$

The second law of thermodynamics requires that

$$q\cdot\operatorname{grad}\theta \leqslant 0, \tag{3.5}$$

and we assume that this inequality is strict for $\operatorname{grad}\theta \neq 0$. We impose the boundary conditions

$$\left.\begin{aligned} &v\cdot n\big|_{\partial\omega} = 0, \\ &\theta\big|_{\partial\omega_2} = \theta_0, \quad q\cdot n\big|_{\partial\omega\setminus\partial\omega_2} = 0 \end{aligned}\right\} \tag{3.6}$$

where $\partial\omega_2$ is a nonempty subset of the boundary $\partial\omega$, $n = n(x)$ is the unit outward normal to $\partial\omega$ at x, and $\theta_0 > 0$ is constant.

As in the previous section, solutions of (3.1) - (3.6) satisfy

$$\frac{d}{dt}\int_\omega \rho\left(\frac{1}{2}|v|^2 + U - \theta_0\eta\right)dx = \theta_0\int_\omega \frac{q\cdot\operatorname{grad}\theta}{\theta^2}\,dx \leqslant 0 \tag{3.7}$$

(cf [11]). We also have the mass constraint

$$\int_\omega \rho\,dx = M, \tag{3.8}$$

where the constant $M > 0$ is determined by the initial data.

Corresponding to (3.7) our aim is to study the absolute minimizers and minimizing sequences of

$$E(\rho,v,\theta) \overset{\text{def}}{=} \int_\omega \rho\left(\frac{1}{2}|v|^2 + U(\rho,\theta) - \theta_0\eta(\rho,\theta)\right)dx \tag{3.9}$$

subject to the constraint (3.8). We make the following hypotheses on $A(\rho,\theta)$:

(i) $A : (0,b)\times(0,\infty) \longrightarrow \mathbb{R}$ is continuous, where $b > 0$ is a constant,

(ii) for each fixed $\rho \in (0,b)$, $A(\rho,\cdot)$ is C^1 and strictly concave,

(iii) for each fixed $\theta \in (0,\infty)$, the function $f_\theta(\rho) \overset{\text{def}}{=} \rho A(\rho,\theta)$ satisfies $\lim_{\rho\to 0+} f_\theta(\rho) = 0$, $\lim_{\rho\to 0+} \frac{f_\theta(\rho)}{\rho} = -\infty$ and $\lim_{\rho\to b-} f_\theta(\rho) = +\infty$.

These hypotheses are satisfied by the classical van der Waals' fluid (cf Landau & Lifshitz [23]) for which

$$A(\rho,\theta) = -a\rho + k\theta \log\left(\frac{\rho}{b-\rho}\right) - c\theta\log\theta - d\theta + \text{const.}, \quad (3.10)$$

where the coefficients a,k and c are positive.

By (ii)

$$A(\rho,\theta) \leqslant A(\rho,\theta_o) + (\theta-\theta_o)\frac{\partial A}{\partial\theta}(\rho,\theta_o),$$

with equality if and only if $\theta = \theta_o$. Thus the integrand in (3.9) has a strict minimum, for fixed ρ, when $v = 0$ and $\theta = \theta_o$. Motivated by this, we consider the problem of minimizing

$$I(\rho) \overset{\text{def}}{=} \int_\omega \rho(U(\rho,\theta_o) - \theta_o\eta(\rho,\theta_o))dx$$

$$= \int_\omega f_{\theta_o}(\rho(x))dx$$

among measurable functions $\rho : \omega \longrightarrow [0,b]$ satisfying (3.8), where $f_{\theta_o}(b)$ is defined to be $+\infty$ in consonance with (iii).

We are interested in cases, such as (3.10), for which $f_{\theta_o}(\cdot)$ is not convex. We denote by $f_{\theta_o}^{**}$ the <u>lower convex envelope</u> of f_{θ_o}, that is

$$f_{\theta_o}^{**}(\rho) = \sup\{\alpha + \beta\rho : \alpha + \beta t \leqslant f_{\theta_o}(t) \text{ for all } t \in [0,b)\},$$

and by $\mathscr{W}$ the <u>Weierstrass set</u>

$$\mathscr{W} = \{\rho \in [0,b) : f_{\theta_o}^{**}(\rho) = f_{\theta_o}(\rho)\}.$$

Recall that if $F : [0,b) \longrightarrow \mathbb{R}$ then the subdifferential $\partial F(\rho)$ of F at the point $\rho \in [0,b)$ is defined to be the set

$$\partial F(\rho) \overset{\text{def}}{=} \{\beta \in \mathbb{R} : F(\rho) + \beta(t-\rho) \leqslant F(t) \text{ for all } t \in [0,b)\}.$$

Define

$$S(\bar\rho) = \{\rho \in (0,b) : \partial f_{\theta_o}^{**}(\bar\rho) \subset \partial f_{\theta_o}(\rho)\},$$

where $\overline{\rho} = \frac{M}{\text{meas } \omega}$ is the mean density. It is easily shown that $S(\overline{\rho}) \subset \mathcal{W}$ and, using (iii), that $\overline{\rho}$ belongs to the convex hull of $S(\overline{\rho})$. In the case of (3.10), for $\frac{ab}{k\theta_o} > \left(\frac{3}{2}\right)^3$ there exists exactly one nontrivial common tangent to the graph of f_{θ_o} with end-points ρ_1, ρ_2 as shown in Figure 1. The Weierstrass set $\mathcal{W} = [0,\rho_1] \cup [\rho_2,b)$, and $S(\overline{\rho}) = \{\overline{\rho}\}$ for $\overline{\rho} \in (0,\rho_1) \cup (\rho_2,b)$, $S(\overline{\rho}) = \{\rho_1\} \cup \{\rho_2\}$ for $\overline{\rho} \in [\rho_1,\rho_2]$. In order to characterize the minimizing

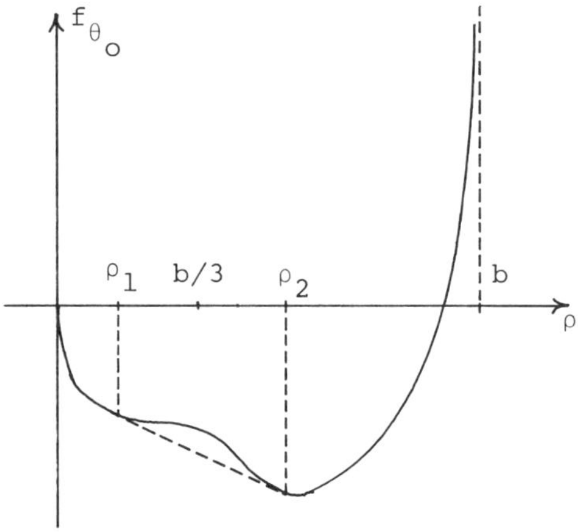

Figure 1

The graph of f_{θ_o} *for a van der Waals' fluid*

sequences of I we introduce, following L.C. Young [30] (see also McShane [24], Berliocchi & Lasry [10], Tartar [27]), the generalized problem:

Minimize

$$\hat{I}(\nu) \stackrel{\text{def}}{=} \int_\omega \int_{[0,b]} f_{\theta_o}(\rho)\, d\nu_x(\rho)\, dx$$

subject to

$$\int_\omega \int_{[0,b]} \rho\, d\nu_x(\rho)\, dx = M. \tag{3.11}$$

The unknown $\nu = (\nu_x)$ is a Young measure, that is a measurable mapping $x \longmapsto \nu_x$ of ω to probability measures on $[0,b]$. (Due to the results of Tartar [27,28] and DiPerna

[14,15] these measures are playing an increasing rôle in the study of nonlinear partial differential equations. Their use in the calculus of variations is now standard; see, for example, the article by Jean Taylor in this volume.) An ordinary function $\rho(x)$ corresponds to the Young measure $\nu_x = \delta_{\rho(x)}$; note that for this ν we have $\hat{I}(\nu) = I(\rho)$ and $\int_\omega \int_{[0,b]} \rho d\nu_x(\rho)dx = \int_\omega \rho(x)dx$.

Theorem 3.1

(a) *The minimum of* $\hat{I}(\nu)$ *subject to (3.11) is attained; the minimizing Young measures* $\bar{\nu}$ *are exactly those satisfying (3.11) and such that* $\mathrm{supp}\,\bar{\nu}_x \subset S(\bar{\rho})$ *a.e.* $x \in \omega$.

(b) *The minimum value of* I *subject to (3.8) is the same as that of* $\hat{I}(\nu)$ *subject to (3.11), and is attained exactly by those functions* ρ *satisfying (3.8) and such that* $\rho(x) \in S(\bar{\rho})$ *a.e.* $x \in \omega$.

(c) *Let* $\{\rho_j\}$ *be any minimizing sequence for* I *subject to (3.8); then there exists a subsequence* $\{\rho_\mu\}$ *and a minimizing Young measure* $\bar{\nu}$ *for* $\hat{I}$ *subject to (3.11), such that for any continuous function* $F : [0,b] \to \mathbb{R}$,

$$F(\rho_\mu) \overset{*}{\longrightarrow} \int_{[0,b]} F(\rho)d\bar{\nu}_x(\rho) \quad \text{in} \quad L^\infty(\omega). \tag{3.12}$$

Conversely, given any minimizing Young measure $\bar{\nu}$ *for* $\hat{I}$ *subject to (3.11) there exists a minimizing sequence* $\{\rho_\mu\}$ *of* I *subject to (3.8) satisfying (3.12).*

Part (b) of the theorem is a result of a type first stated by Gibbs [21]; a similar, but not identical version is given by Dunn & Fosdick [17 Theorem 9].

The proof of Theorem 3.1 is given in [6], where a variety of similar problems are also considered. Here we merely note that part (a) follows by integration of the inequality

$$f(\rho) \geq f(\tilde{\rho}) + \beta(\rho - \tilde{\rho}), \ \rho \in [0,b], \ \tilde{\rho} \in S(\bar{\rho}), \ \beta \in \partial f_{\theta_0}^{**}(\bar{\rho}),$$

with respect to $\bar{\nu}_x, \nu_x$ and ω.

Applying part (a) to $f_{\theta_o}^{**}$, and noting that $\partial f_{\theta_o}^{**}(\rho) = \partial f_{\theta_o}(\rho)$ for any $\rho \in \mathcal{W}$, we see that the minimizing Young measures $\bar{\nu}$ of $\hat{I}$ are the same as those of

$$\hat{I}^{**}(\nu) = \int_\omega \int_{[0,b]} f_{\theta_o}^{**}(\rho)\, d\nu_x(\rho)\, dx,$$

and hence that the infimum of I is unchanged if f_{θ_o} is replaced by $f_{\theta_o}^{**}$. It follows using lower semicontinuity that any minimizing sequence of I has a subsequence converging weak $*$ in $L^\infty(\omega)$ to a minimizer of

$$I^{**}(\rho) = \int_\omega f_{\theta_o}^{**}(\rho(x))\, dx$$

subject to (3.8). A result closely related to this was proved by Dacorogna [12] in material coordinates. The appearance of the lower convex envelope of f_{θ_o} is consistent with results of statistical physics for infinite volumes (for discussion and references see Thompson [29]); these results establish convexity properties of certain averaged free energy functions, but do not appear to give information concerning the local free energy A. Note that part (b) of the theorem shows that only values of $\rho \in \mathcal{W}$ can be observed in an absolute minimizer; this is the classical Weierstrass condition of the calculus of variations. Sometimes it is erroneously asserted that because of this 'stability' condition f_{θ_o} is itself convex; the correct interpretation noted above has been pointed out by, for example, Ericksen [18].

Using Theorem 3.1 it can be shown that any minimizing sequence (ρ_j, v_j, θ_j) of E subject to (3.8) possesses a subsequence $(\rho_\mu, v_\mu, \theta_\mu)$ such that $v_\mu \longrightarrow 0$ a.e., $\theta_\mu \longrightarrow \theta_o$ a.e., and ρ_μ converges to a minimizer $\bar{\nu}$ of $\hat{I}$ in the sense of (3.12). Similarly, the minimizers of E have the form $(\rho, 0, \theta_o)$, where ρ is a minimizer of I. It follows from part (c) of the theorem that in general (e.g. for the van der Waals' fluid) there are minimizing sequences (ρ_j, v_j, θ_j) with ρ_j converging in the sense of (3.12) to a Young measure $\bar{\nu}$ which does not correspond to a function, and such that $\rho_j \xrightarrow{*} \rho$ in $L^\infty(\omega)$ with ρ not a minimizer of I. Typically these sequences 'mix the phases' more and more finely as j increases. It would be very interesting to know if such sequences can be realized by the dynamics (cf (Q5) and Andrews & Ball [2]). Such dynamic behaviour would be physically significant in regimes where the phases are mixed sufficiently coarsely to neglect the energy of phase boundaries.

Finally we note that by part (b) of the theorem, any minimizer $(\rho, 0, \theta_o)$ of E satisfies

$$f'_{\theta_o}(\rho(x)) = \text{const.}\ ,\quad f_{\theta_o}(\rho(x)) - \rho(x) f'_{\theta_o}(\rho(x)) = \text{const.}\ ,$$

a.e. in ω, provided $A(\rho,\theta)$ is C^1 in ρ. These are the familiar necessary conditions representing constancy of the chemical potential and pressure respectively.

4. QUASICONVEXITY AND ELASTIC STABILITY

We consider a nonlinear thermoelastic material as in §2. For simplicity we suppose that the material is homogeneous, so that ρ_R and A do not depend explicitly on X. In contrast to §2 we suppose that there is a conservative body force $\rho_R b = \nabla_x \Psi(X,x)$, so that (2.1) now becomes

$$\rho_R \ddot{x} = \operatorname{Div} T_R + \rho_R b. \tag{4.1}$$

We assume that the boundary conditions are as in §2, special case 1, with $\partial\Omega_2$ nonempty. The same calculation as usual shows that if

$$\mathcal{F} = \rho_R\left[\frac{1}{2}|\dot{x}|^2 + U - \theta_o \eta\right] + \Psi$$

then $\frac{d}{dt}\int_\Omega \mathcal{F} dX \leq 0$. Assuming that A is strictly concave in θ and applying the reasoning in §3, we are led to consider the problem of minimizing

$$I(x) = \int_\Omega [W(\nabla x(X)) + \Psi(X, x(X))]\,dX \tag{4.2}$$

subject to (2.5), where the <u>stored-energy function</u> W is defined by

$$W(F) \overset{\text{def}}{=} \rho_R A(F, \theta_o). \tag{4.3}$$

We suppose that $W : M^{n\times n} \longrightarrow \overline{\mathbb{R}}$ is continuous (with respect to the usual topology of the extended real line $\overline{\mathbb{R}}$) and bounded below, where $M^{n\times n}$ denotes the set of all real $n \times n$ matrices, and that $\Psi : \overline{\Omega} \times \mathbb{R}^n \longrightarrow \mathbb{R}$ is continuous and bounded below.

The following definition is an adaptation of that of Morrey [26].

Definition ([9])

Let $1 \leqslant p \leqslant \infty$. W is $W^{1,p}$-quasiconvex at $A \in M^{n\times n}$ if

$$\int_D W(A + \nabla\phi(Y))dY \geqslant \int_D W(A)dY$$

for every bounded open set $D \subset \mathbb{R}^n$ with meas $\partial D = 0$ and all ϕ belonging to the Sobolev space $W_0^{1,p}(D;\mathbb{R}^n)$. If this holds for all $A \in M^{n\times n}$ we say that W is $W^{1,p}$-quasiconvex.

We attempt to illuminate this somewhat impenetrable condition by stating some recent results.

Theorem 4.1 (Ball & Murat [9])

Let $A \in M^{n\times n}$, *and suppose that* $I(x)$ *attains a minimum on* $AX + W_0^{1,p}(\Omega;\mathbb{R}^n)$ *for every smooth nonnegative* Ψ. *Then* W *is* $W^{1,p}$*-quasiconvex at* A.

It is possible that I is sequentially weakly lower semicontinuous on $W^{1,p}(\Omega;\mathbb{R}^n)$ (weak * if $p = \infty$) if and only if W is $W^{1,p}$-quasiconvex but so far only partial results have been obtained (see [9] for the references). Relaxation theorems of the type given in §3 expressed in terms of lower quasiconvex envelopes (but not making use of the Young measure) have been given by Acerbi & Fusco [1] and Dacorogna [13], though these have not as yet been shown to hold under weak enough growth conditions to apply to elasticity.

Definitions

(a) By a standard boundary region with normal $N \in \mathbb{R}^n$ we mean a bounded strongly Lipschitz domain $D \subset \mathbb{R}^n$ satisfying

(i) D is contained in the half-space $K_a^N = \{X \in \mathbb{R}^n : X \cdot N < a\}$ for some $a \in \mathbb{R}^n$, and

(ii) the $n-1$ dimensional interior E of $\partial D \cap K_a^N$ is nonempty; we denote $\partial D \setminus E$ by ∂D_1.

(b) Let $x \in W^{1,1}(\Omega;\mathbb{R}^n)$ be such that $I(x)$ exists and is finite, and let $X_o \in \overline{\Omega}$. We say that x is a <u>local minimum of</u> I <u>at</u> X_o <u>in</u> $W^{r,p} \cap C^o$ if there are numbers $\rho > 0$, $\delta > 0$ such that $I(y)$ exists and $I(y) \geqslant I(x)$ whenever $y - x \in C^{\infty}(\overline{\Omega};\mathbb{R}^n)$, $y(X) = x(X)$ for $|X - X_o| > \rho$ and $X \in \overline{\Omega}$, and $\|y - x\|_{W^{r,p}(\Omega;\mathbb{R}^n)} + \|y - x\|_{C^o(\overline{\Omega};\mathbb{R}^n)} < \delta$.

<u>Theorem 4.2</u> (a special case of Ball & Marsden [7 Thm 2.2])

Let $1 \leqslant p < \infty$ *and let* r *be a positive integer satisfying* $r < 1 + \frac{n}{p}$. *Suppose* $x \in W^{1,1}(\Omega;\mathbb{R}^n)$ *is a local minimum of* I *at* $X_o \in \overline{\Omega}$ *in* $W^{r,p} \cap C^o$ *and that* x *is* C^1 *in a neighbourhood of* X_o *in* $\overline{\Omega}$.

(i) If $X_o \in \Omega$, *then*

$$\int_D W(\nabla x(X_o) + \nabla\phi(Y))\,dY \geqslant \int_D W(\nabla x(X_o))\,dY$$

for any bounded open set $D \subset \mathbb{R}^n$ *and all* $\phi \in C_o^1(D;\mathbb{R}^n)$ *(=* C^1 *functions with compact support in* D*) satisfying* $\det(\nabla x(X_o) + \nabla\phi(Y)) > 0$ *for all* $Y \in \overline{D}$.

(ii) Let $X_o \in \partial\Omega \setminus \partial\overline{\Omega}_1$, *and suppose* $\partial\Omega$ *is smooth in a neighbourhood of* X_o. *Let* $N = N(X_o)$ *be the unit outward normal to* $\partial\Omega$ *at* X_o, *and let* D *be a standard boundary region with normal* N. *Then*

$$\int_D W(\nabla x(X_o) + \nabla\phi(Y))\,dY \geqslant \int_D W(\nabla x(X_o))\,dY$$

for all $\phi \in C^1(\overline{D};\mathbb{R}^n)$ *vanishing in a neighbourhood of* ∂D_1 *in* $\overline{D}$ *and such that* $\det(\nabla x(X_o) + \nabla\phi(Y)) > 0$ *in* $\overline{D}$.

Part (i) of the theorem is but a slight generalization of a result of Meyers [25 pp128-131]; note that the conclusion is nearly that W is $W^{1,\infty}$-quasiconvex at $\nabla x(X_o)$. The condition in part (ii) of the theorem is a <u>quasiconvexity condition at the boundary</u>; roughly, it asserts that $z(Y) = \nabla x(X_o)Y$ minimizes $\int_D W(\nabla z(Y))\,dY$ <u>globally</u> subject to the boundary condition $z|_{\partial D_1} = \nabla x(X_o)Y|_{\partial D_1}$. In [7] part

(ii) is used for $n > 1$ to construct an example of a strictly quasiconvex, strictly polyconvex W having a natural state that is not a local minimum of $J(x) \overset{\text{def}}{=} \int_\Omega W(\nabla x)\,dX$ in $W^{r,p} \cap C^0$ for $r < 1 + \frac{n}{p}$ even though the second variation of J is strictly positive (linearized stability); this cannot happen for $n = 1$. The technical hypotheses in results such as Theorem 4.2 could do with some improvement to allow less regularity of $x(\cdot)$ and $\phi(\cdot)$.

Example (cf [4,9])

Let $n = 3$ and define

$$W(F) = \operatorname{tr}(F^T F) + h(\det F), \tag{4.4}$$

where h is convex, $h(\delta) = +\infty$ for $\delta \leqslant 0$, h is continuous for $\delta > 0$, and $\lim_{\delta \to 0+} h(\delta) = \lim_{\delta \to \infty} \frac{h(\delta)}{\delta} = \infty$. Then W is $W^{1,p}$-quasiconvex if and only if $p \geqslant 3$. In fact if $1 \leqslant p < 3$ then W is not $W^{1,p}$-quasiconvex at $\lambda 1$ for $\lambda > 0$ sufficiently large; this corresponds to the fact that a solid ball B made of this material and subjected to the radial boundary displacement $x(X)\big|_{\partial B} = \lambda X$ can reduce its energy by cavitation, i.e. by forming a hole in its interior. The stored-energy function (4.4) is of a type used to model natural rubber, which can rupture by cavitation.

Given a stored-energy function $W(F)$ one may define for $1 \leqslant p \leqslant \infty$ the sets

$$S_p = \{F : W \text{ is } W^{1,p}\text{-quasiconvex at } F\}.$$

Clearly $S_p \subset S_q$ if $p \leqslant q$. Anticipating the proof of refinements of Theorem 4.2 one can think of S_p as consisting of those F that can be observed in configurations that are local minimizers in $W^{1,p}$. In the example (4.4) we have $S_3 = M^{3\times 3}$, $S_1 \neq M^{3\times 3}$ and can view ∂S_1 as a fracture surface. Note, however, that deformations in which x is discontinuous across a plane do not belong to $W^{1,1}(\Omega;\mathbb{R}^3)$, and therefore that the above framework cannot handle the most common type of fracture; this may not be as serious as it sounds, as there is evidence that in some materials cracks are initiated by cavitation. For another speculative approach to the onset of fracture see Ball & Mizel [8].

As a final result concerning quasiconvexity we mention the recent beautiful theorem of Knops & Stuart [22] which states that if W is strictly $W^{1,\infty}$-quasiconvex and C^1 for $\det F > 0$ then for zero body forces the only smooth solution of the equilibrium equations

$$\frac{\partial}{\partial X^\alpha}\frac{\partial W}{\partial x^i_{,\alpha}} = 0 \quad \text{in } \bar{\Omega}$$

satisfying $\det \nabla x(X) > 0$ in $\bar{\Omega}$ and the homogeneous boundary data

$$x(X)\Big|_{\partial\Omega} = AX$$

is $x(X) \equiv AX$, provided Ω is <u>star-shaped</u>.

REFERENCES

1. Acerbi, E. and N. Fusco, Semicontinuity problems in the calculus of variations, Arch. Rat. Mech. Anal., to appear.

2. Andrews, G. and J.M. Ball, Asymptotic behaviour and changes of phase in one-dimensional viscoelasticity, J. Differential Equations <u>44</u> (1982), 306-341.

3. Ball, J.M., Convexity conditions and existence theorems in nonlinear elasticity, Arch. Rat. Mech. Anal. <u>63</u> (1977), 337-403.

4. Ball, J.M., Discontinuous equilibrium solutions and cavitation in nonlinear elasticity, Phil. Trans. Roy. Soc. London A <u>306</u> (1982), 557-611.

5. Ball, J.M., Minimizers and the Euler-Lagrange equations, Proceedings of ISIMM conference, Paris, Springer-Verlag, to appear.

6. Ball, J.M. and G. Knowles, forthcoming.

7. Ball, J.M. and J.E. Marsden, Quasiconvexity at the boundary, positivity of the second variation and elastic stability, to appear.

8. Ball, J.M. and V.J. Mizel, Singular minimizers for regular one-dimensional problems in the calculus of variations, Bull. Amer. Math. Soc., to appear.

9. Ball, J.M. and F. Murat, $W^{1,p}$-quasiconvexity and variational problems for multiple integrals, to appear.

10. Berliocchi, H. and J.M. Lasry, Intégrandes normales et mesures paramétrées en calcul des variations, Bull. Soc. Math. France 101 (1973), 129-184.

11. Coleman, B.D. and E.H. Dill, On thermodynamics and the stability of motion of materials with memory, Arch. Rat. Mech. Anal. 51 (1973), 1-53.

12. Dacorogna, B., A relaxation theorem and its application to the equilibrium of gases, Arch. Rat. Mech. Anal. 77 (1981), 359-386.

13. Dacorogna, B., Quasiconvexity and relaxation of non convex problems in the calculus of variations, J. Funct. Anal. 46 (1982), 102-118.

14. DiPerna, R.J., Convergence of approximate solutions to conservation laws, Arch. Rat. Mech. Anal. 82 (1983), 27-70.

15. DiPerna, R.J., Convergence of the viscosity method for isentropic gas dynamics, Comm. Math. Phys. 91 (1983), 1-30.

16. Duhem, P., "Traité d'Énergetique ou de Thermodynamique Générale", Gauthier-Villars, Paris, 1911.

17. Dunn, J.E. and R.L. Fosdick, The morphology and stability of material phases, Arch. Rat. Mech. Anal. 74 (1980), 1-99.

18. Ericksen, J.L., Thermoelastic stability, Proc. 5th National Cong. Appl. Mech. (1966), 187-193.

19. Ericksen, J.L., Loading devices and stability of equilibrium, in "Nonlinear Elasticity" pp161-174 ed. R.W. Dickey, Academic Press, New York, 1973.

20. Ericksen, J.L., Equilibrium of bars, J. Elasticity, 5 (1975), 191-201.

21. Gibbs, J.W., Graphical methods in the thermodynamics of fluids, Trans. Connecticut Acad. 2 (1873), 309-342.

22. Knops, R.J. and C.A. Stuart, Quasiconvexity and uniqueness of equilibrium solutions in nonlinear elasticity, to appear.

23. Landau, L.D. and E.M. Lifshitz, "Statistical Physics", Pergamon, Oxford, 1970.

24. McShane, E.J., Relaxed controls and variational problems, SIAM J. Control 5 (1967), 438-485.

25. Meyers, N.G., Quasi-convexity and lower semicontinuity of multiple variational integrals of any order, Trans. Amer. Math. Soc. 119 (1965), 225-249.

26. Morrey, C.B., Quasi-convexity and the lower semicontinuity of multiple integrals, Pacific J. Math. 2 (1952), 25-53.

27. Tartar, L., Compensated compactness and partial differential equations, in "Nonlinear Analysis and Mechanics: Heriot-Watt Symposium Vol. IV" pp136-212, ed. R.J. Knops, Pitman, London, 1979.

28. Tartar, L., The compensated compactness method applied to systems of conservation laws, in "Systems of Nonlinear Partial Differential Equations" pp263-285, ed. J.M. Ball, Reidel, 1983.

29. Thompson, C.J., "Mathematical Statistical Mechanics", Macmillan, New York, 1972.

30. Young, L.C., "Lectures on the calculus of variations and optimal control theory", W.B. Saunders, Philadelphia, 1969.

The author was supported by a U.K. Science and Engineering Research Council Senior Fellowship.

Department of Mathematics
Heriot-Watt University
Edinburgh, Scotland.

MECHANICAL ANALOGS OF COEXISTENT PHASES

E. Chater and J. W. Hutchinson

1. INTRODUCTION

Certain mechanical systems display transitions between two nominally uniform solution states which have certain features in common with true phase transitions. Three such examples will be discussed here. In order, they are the bulging of a long cylindrical balloon, neck propagation along bars of certain polymeric materials, and buckle propagation along externally pressurized pipes. Most of the results presented here were taken from two earlier papers by the authors and a colleague [1, 2].

2. STEADY-STATE INFLATION OF A CYLINDRICAL PARTY BALLOON

Imagine a long party balloon with a long uniform cylindrical section in its mid-region. The properties of most balloon rubbers are such that the pressure-volume relation of a cylindrical slice undergoing a purely cylindrical deformation has the qualitative features shown in Fig. 1. The balloon is treated as a membrane with thickness small compared to radius. A purely cylindrical deformation is defined as a deformation in which the slice is imagined to undergo a uniform expansion of its radius and a uniform axial elongation such that the circumferential and axial stresses,

ISBN 0-12-309770-3

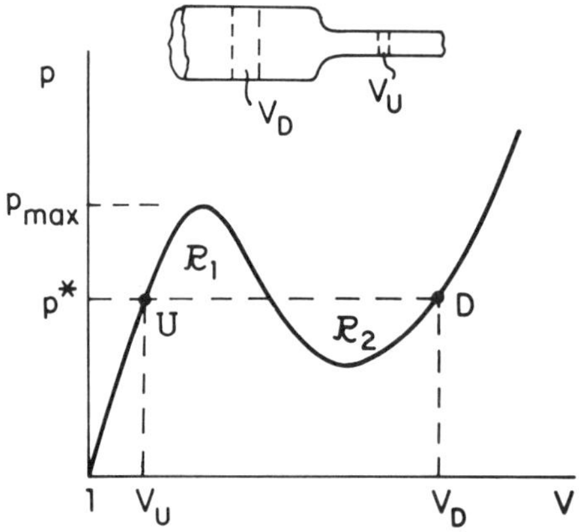

Fig. 1 p(V) for purely cylindrical deformation of a cylindrical segment of unit initial volume. Quasi-static, steady-state propagation requires $\mathcal{R}_1 = \mathcal{R}_2$. (Figure taken from [1].)

respectively, are given by $\sigma_\theta = pR/t$ and $\sigma_x = pR/(2t)$ where R is the current radius, t is the current thickness and p is the internal pressure. The slice considered in Fig. 1 is taken to have a unit volume in the undeformed state. For definiteness it will be assumed that the balloon is inflated under isothermal conditions, and the purely cylindrical deformation in Fig. 1 should also be regarded as isothermal.

The relation of pressure to change of volume of the entire balloon during the inflation process is depicted in Fig. 2. As air is forced into the balloon, a localized bulge forms somewhere along the length of the balloon, usually at one of the ends. The pressure peaks with the initial bulge formation. With continued inflation the pressure settles down to a constant value, p* in Fig. 2, and during this part of the process the transition front between the bulged and unbulged regions simply translates down the length of the balloon with essentially no change in radii of the regions on either side of the transition. This is the portion of the inflation process we will refer to as steady-state propagation. If the balloon is inflated slowly, as is

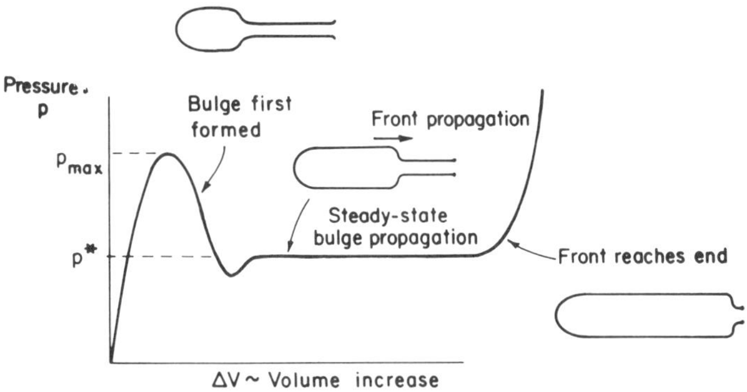

Fig. 2 Inflation of a cylindrical party balloon. (Figure taken from [1].)

assumed to be the case, inertial effects are negligible and the propagation is quasi-static. An example of a partially inflated party balloon is shown in Fig. 3. When the transition front has engulfed the whole balloon the pressure rises and the mid-region again undergoes essentially purely cylindrical deformations.

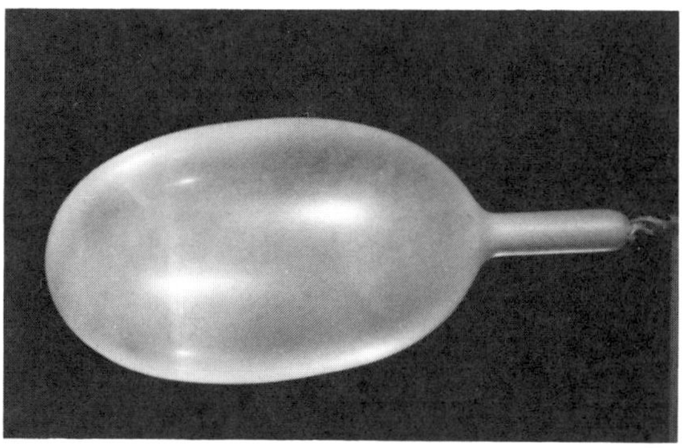

Fig. 3 Party balloon showing transition between bulged and unbulged sections. (Figure taken from [1].)

The equation for the steady-state, quasi-static propagation pressure p^* is obtained by a very simple energy balance argument. Namely, the work done by p^* must equal the change of strain energy stored in the balloon in a unit advance of the transition front. Model the mid-region of the balloon by an infinitely long balloon with uniform properties. Let V_D and V_U denote the volumes of cylindrical sections, each with unit undeformed volume, associated with purely cylindrical deformation states U and D far ahead and far behind, respectively, the transition. These states are each associated with p^* as indicated in Fig. 1. Under steady-state conditions in which the front engulfs a new section with unit undeformed volume, the work done by p^* is exactly $p^*(V_D-V_U)$ since the shape of the transition does not change. With W denoting the isothermal strain-energy per unit undeformed volume of a cylindrical section, the pressure work must equal W_D-W_U since the strain energy stored in the transition does not change under a steady-state advance of the front.

The deformation states in the transition are not purely cylindrical. Nevertheless, because the rubber is characterized by an energy function, W_D-W_U can be calculated using any deformation history which connects states U and D. Thus, if $p(V)$ denotes the relation depicted in Fig. 1 for purely cylindrical deformations, the strain energy difference equals the work in deforming the section from U to D through this deformation history. That is,

$$W_D - W_U = \int_{V_U}^{V_D} p(V)\,dV \tag{2.1}$$

The equation for p^* is therefore

$$p^*(V_D-V_U) = \int_{V_U}^{V_D} p(V)\,dV \tag{2.2}$$

with the well-known graphical solution requiring equality of the areas of the two lobes, $\mathscr{R}_1$ and $\mathscr{R}_2$, as indicated in Fig. 1.

The above derivation for steady-state propagation along the infinitely long balloon obviously applies whether or not the transition is advancing. The derivation can be reinterpreted as the invariance with respect to an arbitrary shift of the solution in the axial direction. In the terminology of phase transformations [3], (2.2) is the condition for the coexistence of two "phases", D and U, of the infinitely long balloon. The pressure p^* for coexistence is below the peak pressure needed to first form a bulge. For the rubber material analyzed in detail in [1], the initial bulging pressure is about twice p^*. This barrier to the formation of a transition is typical of each analog discussed here.

Yin [4] has given a rather complete and general analysis of the deformation of cylindrical membranes subject to internal pressure. We will draw from his work to show how (2.2) emerges from a direct integration of the equations governing axisymmetric deformations of a cylindrical membrane.

Consider a uniform long circular cylindrical membrane of an incompressible rubber-like material which is capped at its ends. The undeformed radius of the membrane is ρ. Attention is restricted to axisymmetric deformations due to internal pressure p. Let $w(\lambda_1, \lambda_2)$ denote the strain energy function of the rubber per unit undeformed area, where λ_1 and λ_2 are the meridional and azimuthal stretches. Based on earlier work of Pipkin [5], Yin has shown that the two equations of equilibrium can be reduced to the following two equations governing the deflection of the membrane:

$$\lambda_1 \frac{\partial w}{\partial \lambda_1} - w = \text{constant} \tag{2.3}$$

and

$$t_1 \cos \omega = \frac{1}{2} \lambda_2 \rho p \tag{2.4}$$

where $t_1 = \lambda_2^{-1}\partial w/\partial\lambda_1$ is the force per unit length of deformed membrane in the meridional direction and ω is the angle made by the meridional tangent with the axis of symmetry.

Using (2.4) to eliminate $\partial w/\partial\lambda_1$ in (2.3), one can readily show that (2.3) can be re-expressed as

$$pV \cos^{-2}\omega - W = \text{constant} \tag{2.5}$$

where, as before, W is the strain energy per unit undeformed volume of a meridional slice and V is the deformed volume of the same slice.

The constant can be evaluated using state U behind the transition for which $\omega = 0$, so that everywhere along the membrane

$$pV \cos^{-2}\omega - W = pV_U - W_U \tag{2.6}$$

In particular, on the other side of the transition in state D where ω again vanishes, (2.6) becomes

$$pV_D - W_D = pV_U - W_U \tag{2.7}$$

which is equivalent to (2.1) and (2.2).

3. NECK PROPAGATION ALONG CYLINDRICAL TENSILE SPECIMENS OF CERTAIN POLYMERIC MATERIALS

Figure 4, taken from the paper by G'Sell, Aly-Helal and Jonas [6], shows a sequence of pictures of the same tensile specimen taken over a progression of overall elongations. The specimen is a solid circular cylinder of high density polyethylene which has been tested in tension in a relatively stiff testing machine. The machine effectively imposes a constant rate of relative separation of the specimen ends. The load carried by the specimen is measured by a load cell (it is not prescribed). Although it may not be noticeable in the first picture of the sequence, a very slight reduction in cross-section has been introduced by machining at the central section of the specimen to induce the neck to set in near the center of the specimen.

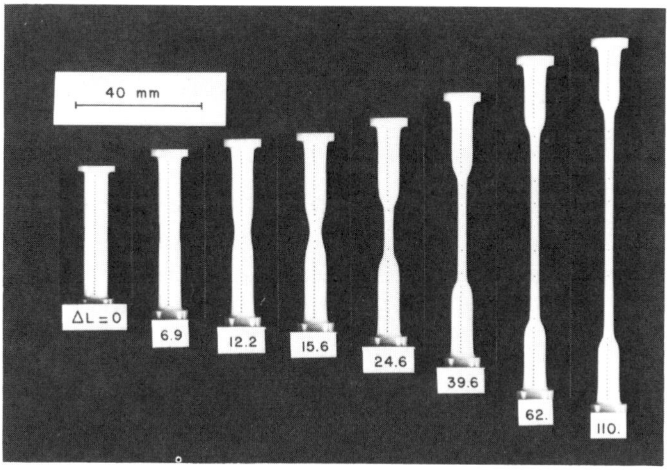

Fig. 4 A sequence of pictures of a solid cylindrical tensile specimen of high density polyethylene displaying neck propagation. (Figure taken from [6].)

The initial stages of neck formation shown in Fig. 4 are very similar to those observed in metal specimens. Significant necking becomes noticeable just following the peak in the overall load-elongation record, and the neck deepens and remains localized as the load continues to fall. In metals, this process continues with monotonically decreasing load until fracture processes interrupt the necking and the specimen fails. Certain polymer specimens, such as that in Fig. 4, propagate the neck once it has become fully localized. It is this aspect which we focus on here. G'Sell et al. [6] have published overall load-elongation records for their tests. These records are qualitatively similar to the overall pressure-volume curve for the cylindrical balloon previously discussed in Fig. 2. After a brief "transient" the neck transition attains a fixed shape and moves along the specimen at a constant velocity, assuming the overall elongation-rate is held constant. Under these steady-state propagation conditions the load is constant, and the radii of the uniform sections of the specimen on either side of the transition do not change. The transition itself extends over an axial distance which is roughly equal to one diameter of the unnecked section.

Constitutive behavior of polymeric materials is not simple. Compared to metals, they have stronger thermal-mechanical coupling and a relatively stronger dependence on the rate of deformation. Moreover, like metals, their multi-axial stress-strain behavior is strongly path-dependent, even when rate-dependency is ignored. Nevertheless, it is very useful to consider a model rubber-like material (i.e., an incompressible, Green-elastic material) whose uniaxial stress-strain curve coincides with that displayed by the polymer at the representative rate of straining. The reason for this is that neck propagation is primarily a consequence of the qualitative shape of the uniaxial stress-strain curve of the material, as will be seen below. The analog between neck propagation in polymers and phase transitions was apparently noted as early as the late 1950's by Thompson and Tuckett (cf. discussion of the paper by Barenblatt [7]). Conditions for the coexistence of necked and unnecked states in a bar subject to uniaxial tension have been considered more recently by Ericksen [8] and James [3]. Here we will review the condition for steady-state propagation assuming the material is nonlinearly elastic and then discuss departures from such ideal behavior when the material is not elastic using results drawn from [2].

Consider a model incompressible, nonlinearly elastic material whose stress-strain behavior in uniaxial tension under isothermal conditions has the features shown in Fig. 5. Here two pairs of work conjugate variables have been used to display the tensile response. While the true stress-log strain curve may be monotonically increasing, the curve of nominal stress (force/original area) versus stretch is assumed to have a local maximum, a local minimum, and then increase monotonicaly to nominal stress levels well above

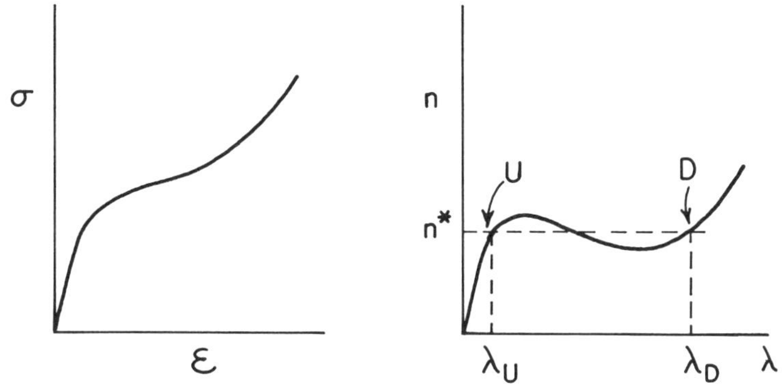

Fig. 5 Stress-strain data in uniaxial tension. True stress versus logarithmic strain on the left and nominal stress (load/original area) versus stretch on the right.

that at the local maximum. The idealized problem is considered for isothermal, steady-state neck propagation along an infinite uniform bar of this material. In the steady-state problem the transition between necked and unnecked regions translates with no shape change towards the unnecked region. An energy balance argument identical to that for the balloon leads to the equation

$$n^*(\lambda_D-\lambda_U) = W_D - W_U \tag{3.1}$$

connecting the nominal stress for quasi-static propagation n^* (i.e., the load per original cross-sectional area) with the stretches in the necked (λ_D) and unnecked (λ_U) regions. Here W is the (isothermal) strain energy density of the material and W_D and W_U denote its values in the uniaxial states far ahead and far behind the transition.

The states of stress in the transition are not uniaxial. Nevertheless, the existence of the strain energy density function W permits us to evaluate W_D-W_U using the uniaxial history $n(\lambda)$ to deform from state U to state D. Since the energy density difference is the work in deforming from U to D , it follows that

$$W_D - W_U = \int_{\lambda_U}^{\lambda_D} n(\lambda)\,d\lambda \tag{3.2}$$

The graphical Maxwell-line solution based on (3.1) and (3.2) is indicated in Fig. 5. It has been tacitly assumed that the solution in the transition is smooth and consistent with quasi-static deformation. This places certain restrictions on W which have not been fully documented. Some polymeric materials whose true stress-strain data in tension has a sharp local maximum exhibit nonsmooth behavior in the form of shear bands, analogous to Lüders bands. But polymeric materials whose true stress-strain curve is monotonically increasing do not appear to give rise to any purely material instabilities, such as shear bands, in neck propagation and the neck transition is smooth.

Some sense of how accurately neck propagation in the nonlinearly elastic model material mimics necking of a more complicated polymeric material was obtained in [2]. In that paper the steady-state problem was formulated and solved approximately for an initially uniform, solid circular cylindrical bar. Two aspects of constitutive behavior departing from nonlinear elasticity were addressed: inherent path-dependence under multiaxial stressing histories, and rate-dependence. Each of these features invalidates the assumptions leading to (3.1) and (3.2) since the strain energy density function W no longer exists. Equation (3.1) continues to hold precisely if $W_D - W_U$ is interpreted as the stress work experienced by a transverse slice of unit volume as it is engulfed by the transition and if n^* is the nominal axial stress averaged across the cross-section (i.e., load/original area, P/A_0). But this stress work difference can no longer be evaluated in terms of the uniaxial history. Now, even to determine quantities such as λ_D , λ_U and

P/A_0 associated with the steady-state solution far ahead and behind the transition, it is necessary to solve the entire problem, including the behavior in the transition.

In [2] the steady-state problem for a solid circular cylindrical bar was formulated as an axisymmetric 3-D flow problem in which the free-surface of the bar was determined as part of the solution process. Approximate solutions were generated using a variational principle together with a numerically implemented Galerkin-Ritz procedure. Here we will comment only on the results obtained for the history dependent constitutive model, the J_2 flow theory of plastic deformation (i.e., Prandtl-Reuss theory based on the Mises invariant). This is a rate-independent constitutive law which is fully specified after it has been made to coincide with data specifying material behavior in uniaxial tension. Calculations were performed for materials with specific uniaxial stress-strain curves chosen to approximate those measured for actual polymeric materials. The results were compared with the predictions based on (3.1) and (3.2) for the nonlinearly elastic model with precisely the same uniaxial stress-strain curve.

Characteristic of the plastic material is a considerably increased resistance to deformation in multiaxial deformation histories which are nonproportional, compared to the response of the corresponding nonlinearly elastic material. Since deformation histories of material elements passing through the neck are decidedly nonproportional, the plastic material offers more resistance to neck propagation than its non-linearly elastic counterpart. For the examples investigated in [2] the nominal load associated with steady-state propagation was between 10 and 20 percent larger for the plastic bar than the elastic bar. The reduction in cross-section from U to D was generally somewhat greater for the plastic material. The work absorbed per slice of unit volume of material as it passes from far ahead to far behind the transition, $W_D - W_U$, is about 30% greater for the plastic bar than the elastic bar. The solution indicates that the shape of the transition is not too different for the two bar materials. The transition is fairly sharp. Its axial extent is approximately one diameter of the unnecked bar.

4. BUCKLE PROPAGATION ALONG PIPES SUBJECT TO EXTERNAL PRESSURE

This phenomena is of some importance in the design of undersea pipelines against collapse [9, 10], and it is this application which provides the background for the work [1] summarized briefly below.

The buckling pressure of a long thin, circular cylindrical shell (pipe) subject to external pressure p is

$$p_c = \frac{E}{4(1-\nu^2)}\left(\frac{t}{R}\right)^3 \qquad (4.1)$$

assuming the shell buckles in the linearly elastic range. Here E and ν are Young's modulus and Poisson's ratio of the material, which is assumed to be isotropic; t is the pipe thickness; and R is its radius. The buckling mode associated with (4.1) is a ring-like deformation in which each cross-section of the pipe undergoes the same ovalization, i.e., a plane strain ring deformation. An undersea pipeline usually has a ratio of t to R in the range 1/15 to 1/50 and is made of a steel with a yield stress which is sufficiently high such that a perfectly (or nearly perfect) circular pipe does undergo bifurcation from the circular state (i.e., does start to buckle) when the stresses are still in the elastic range. Thus, as long as the pipe is not unduly imperfect or damaged, (4.1) provides a good estimate of the maximum pressure the pipe can safely support.

If, however, the pipe suffers a substantial dent at some point along its length or if it buckles locally due to bending in the laying process, then a propagating buckle can be set into motion which spreads over the entire length of the pipe. A short section of a long pipe which has experienced buckle propagation is shown in Fig. 6. The section was selected to show the plastically collapsed section and the transition to the unbuckled circular section. In a manner very similar to the two phenomena discussed earlier, the buckle propagation quickly settles down to a steady-state in which the transition remains fixed in shape and translates along the pipe at constant velocity, assuming the pressure is held constant [10]. The lowest pressure p^* at which steady-state propagation can occur is that associated with low-velocity, quasi-static propagation corresponding to negligible inertial effects. Moreover, the quasi-static propagation pressure p^* can be as little as .2 or even .1 of the "classical" buckling pressure p_c for typical pipe dimensions and materials. Thus a pipe which has suffered severe local damage is susceptible to collapse over its entire length at pressures well below what would normally be considered the buckling pressure.

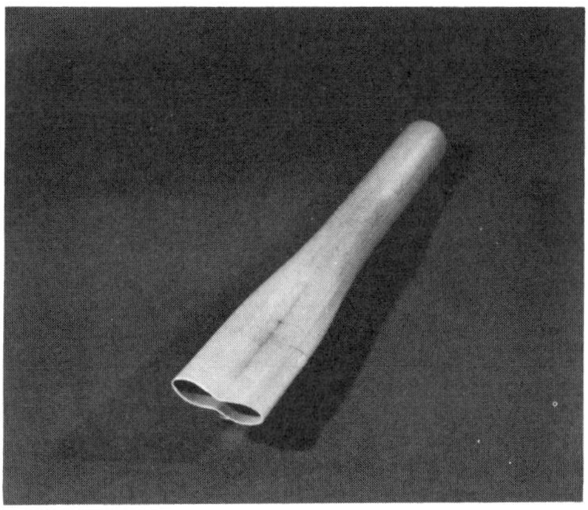

Fig. 6 Section of a pipe showing the transition between the buckled and unbuckled regions of the pipe (pipe section supplied by S. Kyriakides).

The approach put forward in [1] for predicting p^* made use of the buckling and post-buckling solution for a pipe undergoing plane strain ring deformations. The key assumption, or approximation, in developing the model of the pipe is the representation of the pipe material by a nonlinearly elastic material (the deformation theory of plasticity) whose uniaxial stress-strain curve coincides with that of the actual material. This step, which ignores the path-dependence inherent to plastic flow, is analogous (but less drastic) to modeling a polymer by a nonlinearly elastic material, as discussed in the previous section. Invoking a nonlinearly elastic material permits us to connect the ring deformation states far ahead and far behind the transition using the same argument employed in the other two examples. In this way the extremely difficult problem governing behavior in the transition can be side-stepped.

A schematic plot of external pressure p as a function of cross-sectional area decrease ΔA is shown in Fig. 7 for an infinitely long circular cylindrical thin shell undergoing plane strain ring deformation. The yield stress of the material is such that bifurcation (the start of buckling) from the circular state occurs within the elastic range, as already discussed. As ovalization proceeds under slightly increasing pressure, plastic yielding (i.e., nonlinear elastic effects for the nonlinearly elastic material model) begins and a dramatic drop in pressure carrying capacity accompanies further ovalization. When the area decrease ΔA attains approximately 3/4 of the original cross-sectional area, opposite sides of the shell touch and provide an immediate bracing effect. Thereafter, the pressure rises steeply with relatively small additional area decrease.

The state U of the pipe well ahead of the transition is circular and well within the linear elastic range. The collapsed state D far behind the transition is a collapsed ring state. The energy balance argument for propagation of the buckle in the pipe of nonlinearly elastic material under quasi-static, steady-state conditions leads to

$$p^*(\Delta A_D - \Delta A_U) = \int_{\Delta A_U}^{\Delta A_D} p(\Delta A)\, d\Delta A \tag{4.2}$$

Here $p(\Delta A)$ denotes relation between pressure and change of area for plane strain ring deformations so that the Maxwell-line construction for p^* applies, as indicated in Fig. 7.

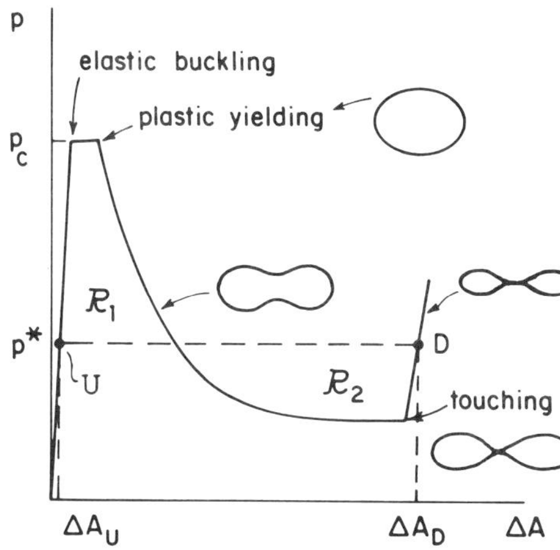

Fig. 7 Buckling and post-buckling of a ring of deformation theory material undergoing plane strain deformation. Schematic curve taken from [1].

Curves of $p(\Delta A)$ calculated using actual uniaxial stress-strain data are given in [1] along with the calculated values of p^*. Comparisons of the predicted values of p^* with experimentally measured quasi-static propagation pressures by Kyriakides are also given in [1] and generally excellent agreement was found. In every case, the theoretical estimate of p^* underestimated the actual propagation pressure, although in most instances only by a few percent. Qualitative arguments involving technical details of plasticity theory can be made to explain why the present simple theory should underestimate measured values of p^* and, additionally, why path-dependent flow effects in the transition area are not of major importance.

REFERENCES

1. Chater, E. and J. W. Hutchinson, On the propagation of bulges and buckles, J. Appl. Mech. (to appear - 1984).
2. Hutchinson, J. W. and K. W. Neale, Neck propagation, J. Mech. Phys. Solids 31 (1983), 405-426.
3. James, R. D., Co-existent phases in one-dimensional static theory of elastic bars, Arch. Rat. Mech. Anal. 72 (1979), 99-140.
4. Yin, W.-L., Non-uniform inflation of a cylindrical elastic membrane and direct determination of the strain energy function, J. Elasticity 7 (1977), 265-282.
5. Pipkin, A. C., Integration of an equation in membrane theory, Z. angen. Math. Phys. 19 (1968), 818.
6. G'Sell, C., N. A. Aly-Helal and J. J. Jonas, Effect of stress triaxiality on neck propagation during the tensile stretching of solid polymers, J. Mater. Sci. 18 (1983), 1731.
7. Barenblatt, G. I., Methods of combustion theory in the mechanics of deformation, flow, and fracture of polymers, in Deformation and Fracture of High Polymers (H. H. Rausch, J. A. Hassell and R. J. Jaffee, eds.), Plenum Press, 1974, 91-111.
8. Ericksen, J. L., Equilibrium of bars, J. Elasticity 5, (1975), 191-201.
9. Palmer, A. C. and J. H. Martin, Buckle propagation in submarine pipelines, Nature 254 (1975), 46-48.
10. Kyriakides, S. and C. D. Babcock, Experimental determination of the propagation pressure of circular pipes, ASME J. Pressure Vessel Tech. 103 (1981), 328-336.

The work of E.C. was supported in part by the Natural Sciences and Engineering Research Council of Canada and by the National Science Foundation under Grant MEA-82-13925. The work of J.W.H. was supported in part by the National Science Foundation under Grant MEA-82-13925 and by the Division of Applied Sciences, Harvard University. C. G'Sell supplied the photograph in Fig. 4.

Division of Applied Sciences
Harvard University
Cambridge, Massachusetts 02138

ADAPTIVE FINITE ELEMENT METHODS AND THE NUMERICAL SOLUTION OF SHEAR BAND PROBLEMS

D. A. Drew and J. E. Flaherty

1. INTRODUCTION.

Shear bands are localized regions of very high shear strain which arise as a result of high rates of loading. They occur in metal forming and cutting processes and in impact and penetration problems. In this paper, we describe a model for the formation of shear bands in simple shear that involves the description of irreversible mechanical shear and the resulting heat release. The location of a shear band is unknown in advance, and the evolution results in large gradients of displacement, velocity, and temperature. Shear band formation, therefore, offers an interesting and physically important application of a code able to resolve small-scale transient structures.

In this paper, we use an adaptive finite element code to solve several problems involving shear band formation. The code automatically locates regions with large gradients and adaptively concentrates finite elements there in order to minimize approximately the discretization error per time step. Our results show the development of shear bands under many circumstances and indicate some possible mechanisms for their formation.

Many technological situations involve the rapid formation, evolution and propagation, and disintegration of small scale structures. Examples include shock waves, shear

ISBN 0-12-309770-3

layers in laminar and turbulent flows, phase boundaries during nonequilibrium thermal processes, shear bands in solid material undergoing rapid strain, and classical boundary layers. With increasing complexity of the physical situations, there is an increasing need for reliable and robust software tools to accurately and efficiently describe the phenomena.

The development of adaptive methods for partial differential equations is an active area of research and a survey of some recent results may be found in Babuska, Chandra, and Flaherty [1]. Techniques divide roughly into two classes: (i) moving grid methods and (ii) local refinement methods. In the former, a fixed number of computational cells or finite elements are moved so as to follow and resolve local nonuniformities such as shock waves, boundary layers, combustion fronts, etc. In the latter, uniform fine grids are added to an underlying coarse grid in regions where the discretization error is estimated to be too large. Moving mesh methods are easier to implement than local refinement methods, which often require the use of relatively sophisticated data structures. They are also very effective at reducing dispersive errors in the vicinity of wave fronts (cf. Hedstrom and Rodrique [12]). However, local refinement methods can, in principle, add enough fine grids to satisfy a prescribed error tolerance, whereas this is generally not possible with methods that use a fixed number of computational cells.

We are studying adaptive finite element techniques for finding numerical solutions of M-dimensional vector systems of partial differential equations that have the form

$$Lu := u_t + f(x,t,u,u_x) - [D(x,t,u)u_x]_x = 0,$$

$$a < x < b, \qquad t > 0 \tag{1.1}$$

subject to the initial and linear separated boundary conditions

$$u(x,0) = u_0(x), \quad a \leqslant x \leqslant b, \tag{1.2a}$$

$$B_1u(a,t) := A_{11}(t)u(a,t) + A_{12}(t)u_x(a,t) = b_1(t), \tag{1.2b}$$

$$B_2u(b,t) := A_{21}(t)u(b,t) + A_{22}(t)u_x(b,t) = b_2(t),$$
$$t > 0. \qquad (1.2c)$$

There are k_1 boundary conditions (1.2b) at $x = a$ and k_2 boundary conditions (1.2c) at $x = b$. We are primarily concerned with parbolic problems where D is positive definite and $k_1 = k_2 = M$; however, we do not restrict ourselves to this case, but instead we assume that conditions are specified so that equations (1.1) and (1.2) have an isolated solution.

Our ultimate goal is to create reliable and robust software that will solve a wide class of problems without requiring users to supply numerical data such as temporal and spatial step sizes. Thus, we envison a computer code that will automatically discretize and adaptively solve (1.1), (1.2) on a nonuniform computational net and attempt to meet a prescribed error tolerance. However, we are still very far from achieving this goal, and herein we describe and use a moving mesh finite element-Galerkin method to discretize (1.1), (1.2) on a grid of trapezoidal space-time elements. This technique is similar to one proposed by Jamet and Bonnerot [13] for fluid flow problems and a detailed dis-discussion and analysis of it was given in Davis [7] and Davis and Flaherty [8]. We are also developing a local refinement procedure for discretizing and solving (1.1), (1.2) and this technique will be described in a forthcoming paper by Flaherty and Moore [11].

The trapezoidal space-time grid is moved so as to approximately minimize the discretization error per time step. This task is known (cf. de Boor [9] or Pereyra and Sewell [14]) to be asymptotically equivalent to selecting a mesh that equidistributes the error, i.e. a mesh where the error is equal on every element. Flaherty, Coyle, Ludwig, and Davis [10] analyzed several mesh equidistribution algorithms and, unfortunately, they show that a great many of them lead to unstable differential equations for the mesh velocities when the partial differential system (1.1), (1.2) is dissipative. Since we are primarily interested in solving parabolic problems, some care has to be exercized when adaptively moving the mesh.

In Section 2 of this paper we construct a simple one-dimensional model that exhibits shear band formation. We simplify this model by approximating the energy equation by a simple heat condution equation. In Section 3 we briefly describe our adaptive finite element code. In Section 4, we present numerical results of several problems that exhibit shear band formation, and, finally, in Section 5 we discuss our results and suggest some future research.

2. SHEAR BAND MODEL.

Consider the one-dimensional shearing motion of a slab of material assumed to lie between $x^* = 0$ and $x^* = L$. The displacement at time t^* of a plane with location x^* is $u^*(x^*,t^*)\underline{j}$, where $\underline{j}$ is a unit vector parallel to the x^*-plane, giving the direction of motion (cf. Figure 1).

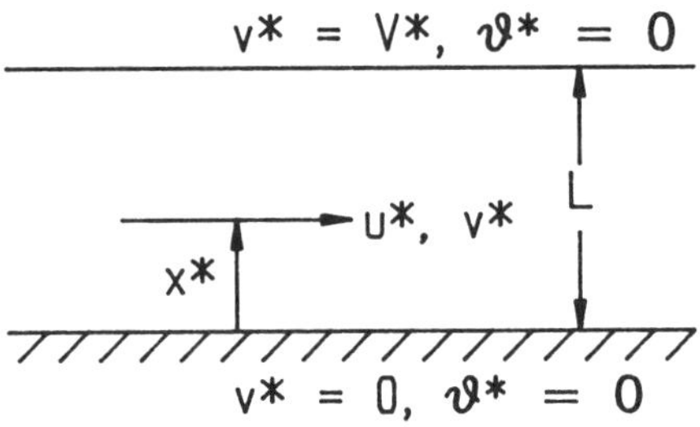

Figure 1: Slab geometry and nomenclature.

The equation of conservation of mass implies that the density $\rho^*(x^*,t^*)$ is independent of t^*, and we further assume that $\rho^*(x^*,t^*) = \rho$, a constant. The momentum equation in the $\underline{j}$ direction is

$$\rho \frac{\partial v^*}{\partial t^*} = \frac{\partial \tau^*}{\partial x^*} , \quad \frac{\partial u^*}{\partial t^*} = v^* , \tag{2.1a,b}$$

where v^* is the velocity and τ^* is the x-y component of the stress tensor, which we shall call the shear stress, or simply stress. The energy equation is

$$\rho \frac{\partial(\varepsilon^* + \frac{1}{2}(v^*)^2)}{\partial t^*} = -\frac{\partial q^*}{\partial x^*} + \frac{\partial \tau^* v^*}{\partial x^*} \tag{2.2}$$

where ε^* is the internal energy and q^* is the heat flux.

Constitutive equations are needed for ε^*, q^* and τ^*. First, we assume

$$q^* = -k^*(\theta^*)\frac{\partial \theta^*}{\partial x^*} \tag{2.3}$$

where θ^* is the temperature and $k^*(\theta^*)$ is the thermal conductivity. We assume that the stress is composed of two parts, an elastic part, plus a viscous stress, i.e.,

$$\tau^* = \rho \frac{\partial \varepsilon^*}{\partial e^*} + \mu(\theta^*) \frac{\partial v^*}{\partial x^*}, \tag{2.4a}$$

where

$$e^* = \partial u^*/\partial x^* \tag{2.4b}$$

is the strain and $\mu^*(\theta^*)$ is the viscosity. Finally, we assume

$$\varepsilon^* = c\theta^* + \frac{1}{2}(\varepsilon^*)^2 G^*(\theta^*) \tag{2.5}$$

where c is the specific heat, and $G^*(\theta^*)$ is the shear modulus. Note that the elastic, or recoverable stress is $\tau_e^* = G^*(\theta^*)e^*$.

We wish to find solutions of eqs. (2.1)-(2.5) for $0 < x^* < L$, and for $t^* > 0$ and, to this end, we prescribe the following initial boundary conditions:

$$\theta^*(x^*,0) = u^*(x^*,0) = v^*(x^*,0) = 0, \quad 0 \leqslant x^* \leqslant L, \tag{2.6}$$

$$\begin{aligned} &\theta^*(0,t^*) = \theta^*(L,t^*) = 0, \quad u^*(0,t^*) = 0, \\ &u^*(L,t^*) = \int_0^{t^*} V^*(s)ds, \quad v^*(0,t^*) = 0, \\ &\qquad\qquad V^*(L,t^*) = V^*(t^*). \end{aligned} \tag{2.7}$$

The initial conditions (2.6) imply that the slab is initially at rest and at ambient temperature, and that temperatures are measured from ambient. The thermal boundary conditions (2.7a,b) correspond to both ends of the slab being held in contact with a heat reservoir at ambient temperature. Since test specimens are generally produced by cutting out a part of a cylinder, leaving two larger cylindrical sections on the ends (cf. Costin et al. [5]) the assumption of heat reservoirs seems appropriate. Finally, conditions (2.7c-f) imply that a shear is applied to the slab by displacing one end relative to the other with speed $V^*(t^*)$.

We nondimensionalize the above problem by introducing the following scaling:

$$x = x^*/L,\ t = t^*/\left(\sqrt{G^*(0)/\rho}\right),\ u = u^*/L,$$
$$v = v^*/\sqrt{G^*(0)/\rho}\ ,\ \theta = \rho c\theta^*/G^*(0)\ , \tag{2.8}$$

and

$$G(\theta) = G^*(\theta^*)/G^*(0),\ k(\theta) = k(\theta^*)/k^*(0),$$
$$\mu(\theta) = \mu^*(\theta^*)/\mu^*(0)\ . \tag{2.9}$$

Two important dimensionless groups appear. They are the Reynolds number

$$Re = L\sqrt{G^*(0)\rho}/\mu^*(0)\ , \tag{2.10a}$$

which is a measure of the dominance of elastic stresses over viscous stresses, and the Prandtl number

$$Pr = c\ \mu^*(0)/k^*(0)\ , \tag{2.10b}$$

which is a measure of the importance of viscous effects to thermal conduction effects.

Substituting (2.8)-(2.10) into (2.1)-(2.5) we obtain the following scaled system of partial differential equations:

$$\frac{\partial u}{\partial t} = v\ ,\ \frac{\partial u}{\partial x} = e\ , \tag{2.11a,b}$$

$$\frac{\partial v}{\partial t} = \frac{\partial}{\partial x}[G(\theta)e] + \frac{1}{Re}\frac{\partial}{\partial x}\left[\mu(\theta)\frac{\partial v}{\partial x}\right], \tag{2.11c}$$

$$\left[1+\frac{1}{2}G'(\theta)e^2\right]\frac{\partial\theta}{\partial t} = \frac{1}{Pr\ Re}\frac{\partial}{\partial x}\left[k(\theta)\frac{\partial\theta}{\partial x}\right] + \frac{1}{Re}\left(\frac{\partial v}{\partial x}\right)^2\ ,$$
$$0 < x < 1,\quad t > 0\ , \tag{2.11d}$$

where the prime in (2.11d) denotes differentiation with respect to temperature. The dimensionless initial and boundary conditions are

$$u(x,0) = v(x,0) = \theta(x,0) = 0,\quad 0 \leqslant x \leqslant 1 \tag{2.12}$$

$$\theta(0,t) = \theta(1,t) = 0,\ u(0,t) = 0,$$
$$u(1,t) = \int_0^t V(s)ds,\ v(0,t) = 0,\ v(1,t) = V(t)\ . \tag{2.13}$$

It is our desire to investigate the combined effect of heat generation by shear waves and thermal softening. Thus, any model for $G(\theta)$ should represent thermal softening, i.e., $G'(\theta) < 0$ for elevated temperatures. However, the effect of the $G'(\theta)$ term in the energy equation (2.11d) is uncertain, and it is often neglected (cf. Costin et al. [5]). The calculations in Section 4 neglect $G'(\theta)$ in (2.11d) and instead use

$$\frac{\partial \theta}{\partial t} = \frac{1}{\text{Pr Re}} \frac{\partial}{\partial x}\left[k(\theta)\frac{\partial \theta}{\partial x}\right] + \frac{1}{\text{Re}}\left(\frac{\partial v}{\partial x}\right)^2 . \qquad (2.11d')$$

We refer to the model consisting of equations (2.11a-c), (2.11d'), (2.12), and (2.13) as the simplified shear band model.

3. NUMERICAL METHODS.

In this section we briefly describe the essential features of our finite element method and of our mesh moving strategies, respectively. Since these have been discussed in [7], [8], and [10], we only repeat those features of our methods that are necessary for the continuity of this paper.

We discretize problem (1.1), (1.2) on the strip

$$S_n := \{(x,t) \mid a \leq x \leq b,\ t_n \leq t \leq t_{n+1}\} , \qquad (3.1)$$

using a finite element-Galerkin procedure. Hence, we approximate $u(x,t)$ on S_n by $U(x,t) \in \mathcal{U}_K$ and select "test" functions $V(x,t) \in \mathcal{V}_K$, where $\mathcal{U}_K$ and $\mathcal{V}_K$ are K-dimensional spaces of $C^0(S_n)$ functions. We then multiply (1.1) by V^T, replace u by U, integrate over S_n, and integrate the time derivative and diffusive terms by parts to obtain the following marching problem for determining $U(x,t)$ in successive strips S_n, n = 0,1,...:

$$U(x,0) = Pu_0(x),\ a \leq x \leq b,\ n = 0 , \qquad (3.2a)$$

$$F(V,U) := \int_{t_n}^{t_{n+1}} \int_a^b \{-V_t^T U + V^T f(x,t,U,U_x) + V_x^T D(x,t,U)U_x\}dxdt + \int_a^b V^T U dx\Big|_{t_n}^{t_{n+1}}$$

$$- \int_{t_n}^{t_{n+1}} V^T D(x,t,U)U_x dt\Big|_a^b = 0 , \qquad (3.2b)$$

$$V \in \mathcal{V}_K,\ (x,t) \in S_n,\ n \geq 0.$$

Here, P is an interpolation operator on the space $\mathcal{U}_K$ and U must also satisfy any essential (Dirichlet) boundary conditions in (1.2b,c).

In order to select finite element bases for $\mathcal{U}_K$ and $\mathcal{V}_K$ we partition S_n into N trapezoids T_i^n, i = 1,...,N, where T_i^n is the trapezoid with vertices (x_{i-1}^n, t_n), (x_i^{n+1}, t_{n+1}), (x_1^{n+1}, t_{n+1}) $x_{i-1}^{n+1}, t_{n+1})$ (cf. Figure 2).

We write $U \in S_n$ as

$$U(x,t) = \sum_{i=0}^{K} c_i(t)\phi_i(x,t) \ , \tag{3.3}$$

where each $\phi_i(x,t)$ is selected to be nonzero only on $T_{i-1}^n \cup T_i^n$. Specifically, we map each T_i^n in the (x,t)-plane into the rectangle

$$R = \{(\xi,\tau) | -1 \leqslant \xi \leqslant 1, \ 0 \leqslant \tau \leqslant 1\} \tag{3.4}$$

in the (ξ,τ)-plane and, at present, we choose $\phi_j(x,t)$, j = 0,1,...,K, to be either piecewise C^0 linear or piecewise C^1 Hermite cubic polynomials in ξ on T_i^n. We also select $\phi_j(x,t)$, j = 0,1,...,K, as a basis for $\mathcal{V}_K$; thus, the dimension K of $\mathcal{U}_K$ and $\mathcal{V}_K$ is either N or 2N for linear or cubic approximations, respectively.

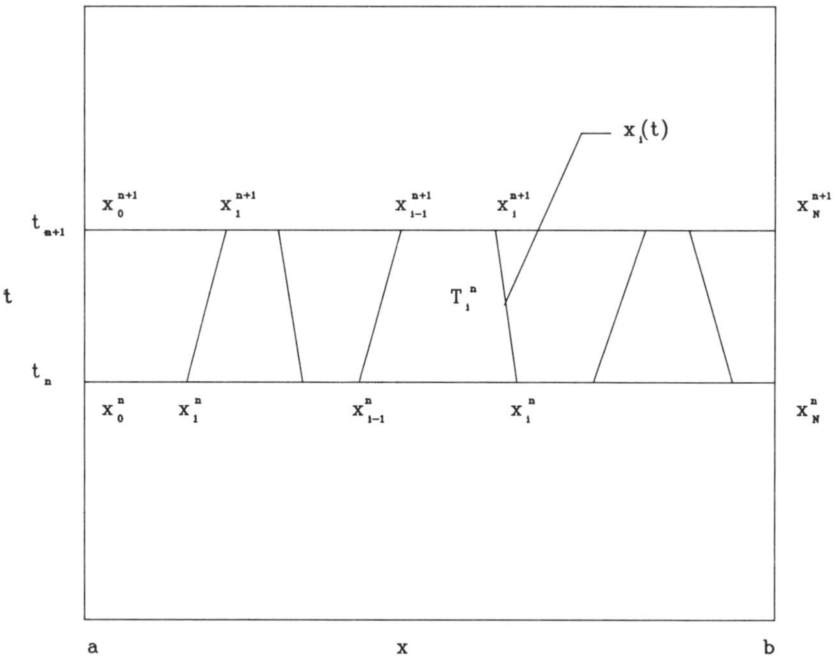

Figure 2: Space-time discretization for the time step $t_n \le t \le t_{n+1}$.

The integrals in equation (3.2b) are transformed element-by-element into integrals over R and are evaluated numerically. We use the Trapezoidal rule to evaluate the integrals and a three-point Gauss-Legendre rule to evaluate the integrals. The resulting system of nonlinear algebraic equations is solved by Newton's method, with users supplying formulas for the Jacobians f_u, f_{u_x}, and D_u.

We now discuss some algorithms for moving the mesh so that the spatial discretization error in L_2 is approximately minimized at each time step. If we assume that the mesh is quasi-uniform, that $u(x,t) \in C^k$ in x for $a < x < b$, and that $\mathcal{U}_k$ is a space of C^0 polynomials of degree $k - 1$, then this task is asymptotically equivalent to equidistributing the local discretization error (cf., e.g., Davis [7], Pereyra and Sewell [14], and Wheeler [16]). Thus, we select the mesh $x_i(t)$, $i = 0,1,\ldots,N$, at time t such that

$$[h_i(t)g(\zeta_i,t)]^k = E(t), \quad i = 1,2,\ldots,N \ , \tag{3.5a}$$

where

$$h_i(t) = x_i(t) - x_{i-1}(t), \tag{3.5b}$$

$$g(x,t)^k = \{[u^{(k)}(x,t)]^T[u^{(k)}(x,t)]\}^{1/2}, \tag{3.5c}$$

$u^{(k)}$ is the kth derivative of u with respect to x, $\zeta_i \in (x_{i-1},x_i)$, $E(t)$ is an undetermined function, and $x_i(t)$ is the line joining x_i^n and x_i^{n+1} (cf. Figure 2).

We solve equations (3.5) for the equidistributing mesh $x_i(t)$, $i = 0,1,\ldots,N$, using a technique developed by de Boor [9] for variable knot spline interpolation. Thus, we begin by taking the kth root of (3.5a) and writing it in an asymptotically equivalent form as

$$\int_{x_{i-1}}^{x_i} g(x,t)dx = c(t) \ , \tag{3.6}$$

where $c(t)^k \approx E(t)$. We let

$$T(x,t) = \int_a^x g(s,t)ds \ . \tag{3.7}$$

Then

$$c(t) = (1/N)T(b,t) \tag{3.8}$$

and the equidistributing mesh $x_i(t)$, $i = 0,1,\ldots,N$, is determined as the solution of the nonlinear system

$$T(x_i,t) = ic(t) \ , \ i = 0,1,\ldots,N \ . \tag{3.9}$$

Of course, $u^{(k)}$ is unknown and it must be approximated by differentiating U. To this end, suppose that we have computed a finite element solution $U(x,t_n)$ at time t_n and on the mesh x_i^n, $i = 0,1,\ldots,N$. We differentiate $U(x,t_n)$ once for piecewise linear approximations or thrice for Hermite cubic approximations and find piecewise constant approximations for $U'(x,t_n)$ or $U'''(x,t_n)$, respectively. We then use five point finite difference approximations of these derivatives to compute values of $U^{(k)}(x_i,t_n)$ and $g(x_i,t_n)$ (cf. (3.5c)) for $i = 0,1,\ldots,N$ and $k = 2$ or 4.

We further assume that $g(x,t_n)$ is a piecewise linear function of x with respect to the mesh x_i^n, $i = 0,1,\ldots,N$, and integrate it to find a piecewise parabolic approximation to $T(x,t_n)$ from (3.7). Finally, we find $c(t_n)$ using (3.8) and determine an approximate equidistributed mesh $\hat{x}_i^n$, $i = 0,1,\ldots,N$, at time t_n by solving (3.9) using the quadratic formula.

The equidistribution algorithm has a non-unique solution whenever $g(x,t) = 0$; therefore, we may expect difficulties whenever $g(x,t)$ is small on any subinterval. We overcome this problem by imposing a lower bound on g, i.e., we replace $g(x,t)$ in equations (3.6) and (3.7) by

$$\tilde{g}(x,t) := g(x,t) + \eta \ , \tag{3.10}$$

where η is a small empirically determined quantity that is discussed further in Davis and Flaherty [8]. Among other things, a positive value of η insures that the solution of (3.9) is a uniform mesh whenever $g(x,t)$ is small everywhere on $[a,b]$.

Our discussion, thus far, has concerned the computation of an equidistribution mesh at time level t_n where a solution $U(x,t_n)$ has already been computed. To obtain an estimate for an optimal mesh at time t_{n+1} prior to computing the solution there, we extrapolate the optimal grids from previous time levels. At the present time, we are using zero order extrapolation, i.e. $x_i^{n+1} = \hat{x}_i^n$, $i = 0,1,\ldots,N$. This strategy has been applied to several examples and, despite its simplicity, it has worked quite well; even on problems with rapidly moving wave fronts. Nevertheless, we can expect that there will be some problems where it will fail to produce an acceptable mesh. However, most of our attempts to use higher order extrapolation produced crisscrossed grids or grids that oscillated wildly from time step-to-time step, even when the solution changed quite little. In order to understand and remedy this phenomenon while simultaneously developing a more dynamic adaptive mesh strategy, we differentiated equation (3.6) with respect to time and obtained the following system for the mesh velocities:

$$\dot{x}_i g(x_i,t) - \dot{x}_{i-1} g(x_{i-1},t) + \int_{x_{i-1}}^{x_i} g_t(x,t)dx = \dot{c}, \qquad i = 1,2,\ldots,N, \tag{3.11}$$

where $\dot{(\ )} := d(\ )/dt$.

Since most higher order (multi-level) mesh extrapolation procedures may be regarded as consistent numerical approximations to equations (3.11), we studied the stability of (3.11) in order to understand some of the difficulties with extrapolation methods. Our analysis is quite general and is neither limited to the specific form of $g(x,t)$ that is given in (3.5c) nor to piecewise linear mesh trajectories.

We assume that $x_i(t)$, $i = 0,1,\ldots,N$, is an equidistributing mesh that exactly satisfies (3.6) and (3.11) and introduce a small perturbation $\delta x_i(0)$, $i = 0,1,\ldots,N$, at $t = 0$. Since x_0 and x_N are fixed, the perturbations must satisfy

$$\delta x_0(t) = \delta x_N(t) = 0 \ , \quad \sum_{i=0}^{N} \delta x_i(t) = 0 \ . \tag{3.12}$$

We assume that no additional errors are introduced; thus, the perturbed system satisfies

$$(\dot{x}_i+\delta\dot{x}_i)g(x_i+\delta x_i,t) - (\dot{x}_{i-1}+\delta\dot{x}_{i-1})g(x_{i-1}+\delta x_{i-1},t) + \int_{x_{i-1}+\delta x_{i-1}}^{x_i+\delta x_i} g_t(x,t)dx = \dot{c} \ , \quad i = 1,2,\ldots,N-1, \ t > 0. \tag{3.13}$$

We further assume $|\delta x_i| \ll 1$, $i = 0,1,\ldots,N$, and linearize (3.13) and find that (3.11) is stable to linear perturbations when

$$L(t) := \max_{1<i<N-1} g(x_i(0),0)/g(x_i(t),t) \tag{3.14}$$

is less than unity (cf. Flaherty et al. [10]). Unfortunately, the choice of $g(x,t)$ given by equation (3.5c), and other reasonable choices, are likely to be decreasing functions of time for dissipative parabolic partial differential equations and this will almost certainly yield a value for $L(t)$ that is larger than one.

Based on a suggestion of Petzold [15], Flaherty et al. [10] showed that the linear combination of equations (3.6) and (3.11) given by

$$\dot{\Phi}_i + \lambda\Phi_i = 0 \ , \ i = 1,2,\ldots,N, \tag{3.15a}$$

where

$$\Phi_i(t) = \int_{x_{i-1}(t)}^{x_i(t)} g(x,t)dx - c(t) \ , \tag{3.15b}$$

is linearly stable provided that the parameter λ satisfies

$$L(t)e^{-\lambda t} < 1 \ . \tag{3.16}$$

Another set of differential equations that approximate (3.6) and give stable mesh dynamics are given by

$$\dot{x}_i - \dot{x}_{i-1} = -\lambda\Phi_i(t) \ , \ i = 1,2,\ldots,N \ . \tag{3.17}$$

Coyle, Flaherty, and Ludwig [6] show that this scheme is linearly stable for all positive values of λ. Of course, conditions (3.16) and (3.17) should be checked locally, i.e., for each step. Furthermore, equations (3.15) or (3.17) will have to be solved numerically and their stability will have to be re-examined in this light. These topics were briefly considered in Flaherty et al. [10] and are analyzed more carefully in Coyle et al. [6].

4. EXAMPLES.

In this section we apply the code that we described in Section 3 to four problems using the simplified shear band model of Section 2. These examples illustrate possible mechanisms for shear band formation.

For all examples we choose

$$k(\theta) = \mu(\theta) = 1 \ , \tag{4.1a,b}$$

$$G(\theta) = (1/2)\left[(1+G_\infty) - (1-G_\infty)\tanh\left(\frac{\theta-\theta_m}{\Delta\theta}\right)\right] \tag{4.1c}$$

Thus, thermal softening occurs in $G(\theta)$ near the temperature θ_m, $\Delta\theta$ measures the scale over which it occurs, and G_∞ is the shear modulus as $\theta \to \infty$. We create a shear in the slab by imposing a nearly constant velocity by prescribing either

$$V(t) = \begin{cases} V_0(t/r) & , \quad 0 < t < r \\ V_0 & , \quad r < t < d-r \\ V_0(t-d)/r & , \quad d-r < t < d \\ 0 & , \quad d < t \end{cases} \qquad (4.2a)$$

or

$$V(t) = V_0 \tanh(t/r) \qquad (4.2b)$$

where V_0 is the shearing velocity, r is the rise time, and d is the duration.

4.1. EXAMPLE 1.

Let us first examine an approximate analytic solution. We assume that $Re >> 1$, $r = 0$ and $d \to \infty$ and, for the moment, that $\theta_m \to \infty$ and $\Delta\theta << 1$ (cf. equation (4.1c)). These simplifications imply that $G(\theta) \approx 1$ and, using equations (2.11a-c) that the displacement approximately satisfies the wave equation

$$\frac{\partial^2 u_0}{\partial t^2} = \frac{\partial^2 u_0}{\partial x^2}, \quad 0 < x < 1, \quad t > 0. \qquad (4.3)$$

We shall study only one pass of an elastic wave, i.e., we find $u_0(x,t)$ for $0 < t < 1$. The solution of equation (4.3) is

$$u_0(x,t) = \begin{cases} 0 & , \quad x + t - 1 < 0 \\ V_0(x + t - 1) & , \quad x + t - 1 > 0 \end{cases}. \qquad (4.4)$$

This solution has a discontinuity in velocity at the wave front, $x + t = 1$, where viscous effects are important. In order to describe the viscous effects, we introduce the "stretched" spatial variable

$$\chi = (x + t - 1)Re^{1/2} \qquad (4.5)$$

into equations (2.11a-c) and find that the inner solution u_i satisfies

$$2\frac{\partial^2 u_i}{\partial t \partial \chi} = \frac{\partial^3 u_i}{\partial \chi^3}, \qquad (4.6)$$

to $O(Re^{-1/2})$. The solution $\hat{u}$ of equation (4.6) that matches to u_0 as $\chi \to \pm\infty$, and is thus a uniform asymptotic approximation to u, is given by

$$\hat{u}_\chi = V_0[1 + \text{erf}(\chi/\sqrt{2t})]/(2\text{Re}^{1/2}) \ . \tag{4.7}$$

Substituting (4.7) into the heat conduction equation (2.11d') while using equations (2.11a-c) and (4.5) and neglecting the thermal conductivity ($\text{Pr} = O(1)$) we find that the temperature approximately satisfies

$$\frac{\partial \theta_0}{\partial t} = \frac{V_0^2}{2\pi t} e^{-2[(x+t-1)\sqrt{\text{Re}/2t}]} \tag{4.8}$$

Since the exponential is only non-negligible near $x + t = 1$, we can further approximate it using the Dirac delta function $\delta(x+t-1)$ and find that

$$\theta_0 = \begin{cases} V_0^2/4\sqrt{2\pi \text{Re}(1-x)} & , \quad x + t < 1 \\ 0 & , \quad x + t > 1 \end{cases} \tag{4.9}$$

The singularity in θ_0 at $x = 1$ could be removed by including the thermal boundary layer near $x = 1$; however, we can see several important features at this point. First, the passage of a (slightly viscous) elastic wave causes significant viscous heating. Note that if we had examined the uniform shear state $u = V_0xt$ (which is not a solution of equations (2.11) for the given initial conditions), we would have obtained $\partial v/\partial x = V_0$, and hence there would be no temperature spike at $x = 1$. It is further plausible that subsequent wave passages will cause additional viscous heating and that, neglecting conduction, this heating will increase the temperature caused by previous wave passages. However, without thermal conduction, the maximum temperature occurs at the edge $x = 1$. We anticipate that by including a small amount of thermal conduction, that the temperature maximum would occur near, but not at, the edge $x = 1$. Finally, we note that viscosity diffuses the elastic wave, and, in so doing, reduces the effect of viscous heating caused by wave passages. For long times, the displacement will approach $u(x,t) = V_0xt$ and the heating will be given by $(\partial v/\partial x)^2/\text{Re} = V_0^2/\text{Re}$ throughout the slab.

We obtained a numerical simulation of the above phenomenon by solving the simplified shear band model with equation (4.2a) and parameter values $\text{Pr}^{-1} = 0.1$, $\text{Re}^{-1} = 0.01$, $\Delta\theta = 0.01$, $\theta_m = 0.03$, $G_\infty = 0.05$, $V_0 = 0.5$, $r = 0.05$, $d \to \infty$.

Solutions for v, θ, and u vs. x for t = 0.05, 0.1, 0.15, 0.2, and 0.25 are shown in Figures 3a, b, and c, respectively. This solution was calculated using 50 finite elements per time step.

If the heating is too small or the conductivity is too large, then no melting occurs during the first passage of the wave. However, it could occur during subsequent passages of reflected waves.

4.2. EXAMPLE 2.

Another mechanism for generating enough heat to cause shear bands is to introduce another wave, e.g., an unloading wave. To illustrate this effect we solve a problem with equation (4.2a) and $Pr^{-1} = 0.2$, $Re^{-1} = 0.05$, $\Delta\theta = 0.01$, $\theta_m = 0.03$, $G_\infty = 0.05$, $V_0 = 0.25$, r = 0.05, d = 1.5. Solutions for v, θ, and u vs. x are shown in Figures 4 a, b, and c, respectively. The stress $\tau(1,t)$ vs. displacement u(1,t) is shown in Figure 4d. The solution was calculated using 50 finite elements per time step and the mesh trajectories that were selected by our adaptive finite element procedure are shown in Figure 5. We see an elastic wave with significant spreading propagating from x = 1 to x = 0, and reflecting. Shortly after reflection, at t = 1.5, the shearing at the right end is stopped. During the propagation of this unloading wave substantial thermal softening occurs near the right end.

As in the previous example, a small piece of the slab near the edge x = 1 is effectively "sheared off". In this example, however, the unloading wave is responsible for causing the temperature spike. As a result, the slab behaves as if it were two independent slabs with essentially zero stress at the shear band (i.e., at $x \approx 0.9$). Subsequently, both parts of the slab oscillate on their own in response to the conditions at the time of shear band formation. The engineering stress-strain curve (cf. Figure 4d) shows a dramatic drop and recovery in stress after the shear band forms. This is presumably due to the elastic propagation in the piece of the slab between $x \approx 0.9$ and x = 1.0.

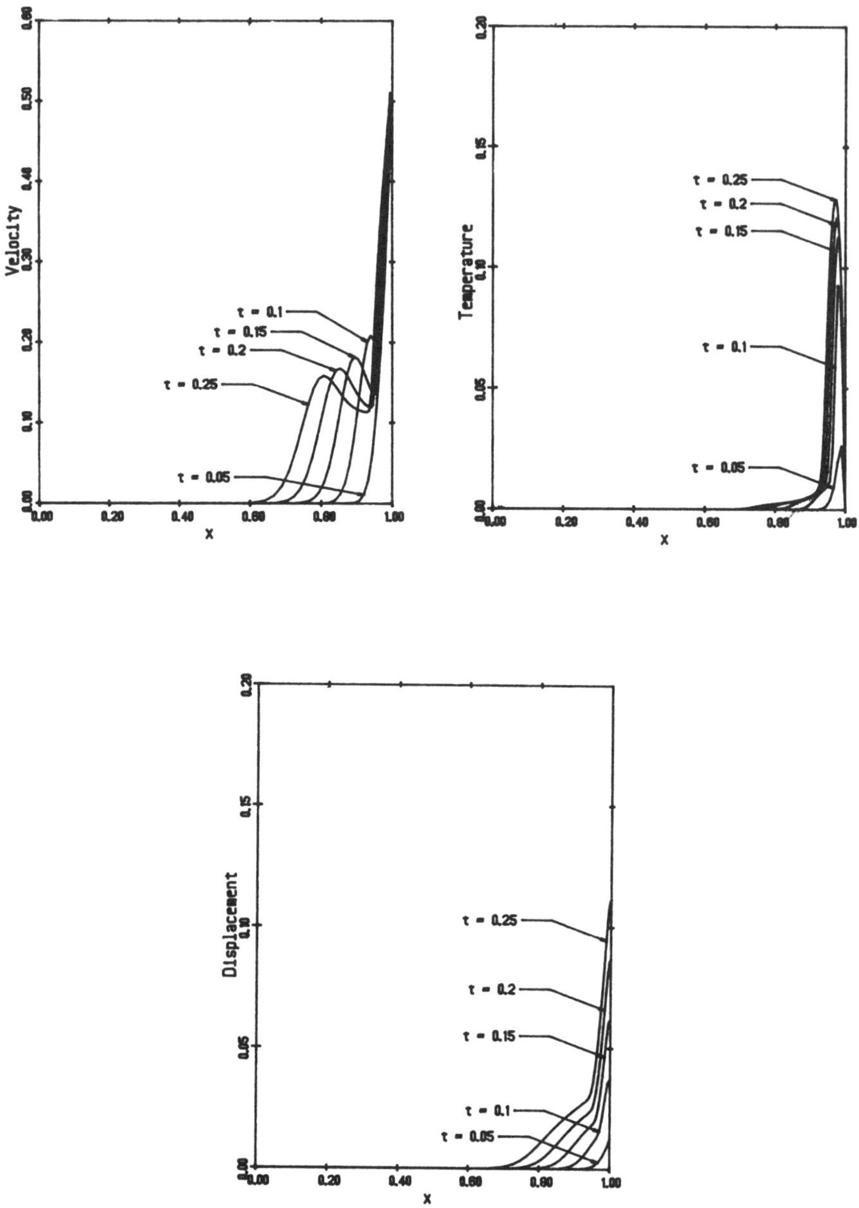

Figure 3: Velocity, temperature, and displacement vs. x at t = 0.05, 0.1, 0.15, 0.2, and 0.25 for Example 1.

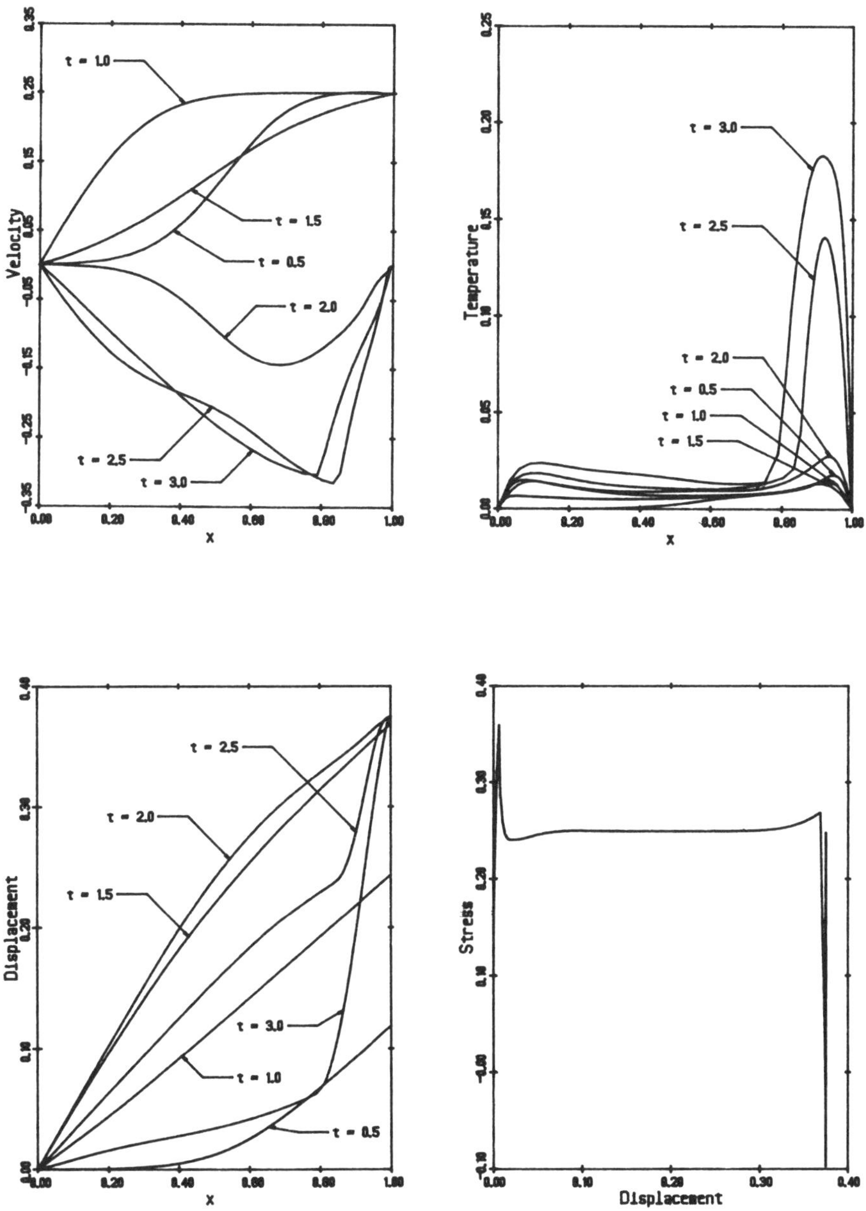

Figure 4: Velocity, temperature, and displacement vs. x at $t = 1.0, 1.5, 2.0, 2.5$, and 3.0 and stress vs. displacement at $x = 1$, $0 \leq t \leq 3.0$ for Example 2.

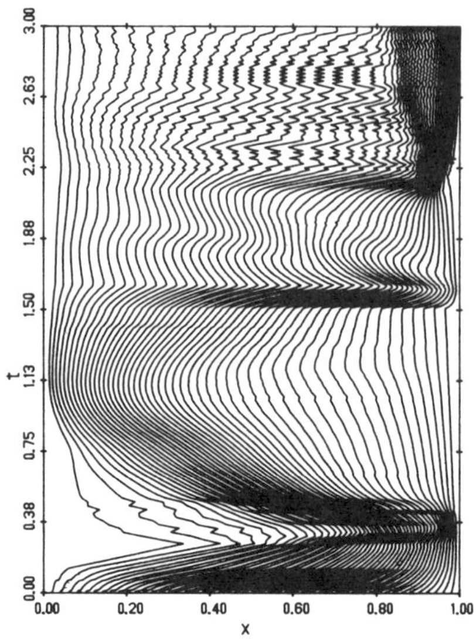

Figure 5: Adaptive mesh trajectories for Example 2.

4.3. EXAMPLE 3.

In this example we study whether or not the overshoot on the engineering stress - displacement diagrams shown in Figure 4d for Example 2 is a numerical artifact caused by the discontinuity in V(t) in equation (4.2a). We do this by solving a problem using the smoother V(t) given by (4.2b) and the parameters of Example 2. Although the overshoot is not as pronounced, the results for the engineering stress vs. displacement shown in Figure 6 imply that the effect is physical and not numerical.

4.4. EXAMPLE 4.

If unlike Examples 1 and 2, the velocity is too small or the conductivity is too large so that the thermal spike near x = 1 is insufficient for thermal softening to occur, the formation of shear bands is still possible, but they will develop according to a different scenario, and at a different location.

In order to see this, suppose that the elastic waves have decayed and the displacement has approached the state of uniform shearing, given by $u(x,t) = V_0xt$. The temperature will evolve toward the steady-state profile $\theta(x)$ =

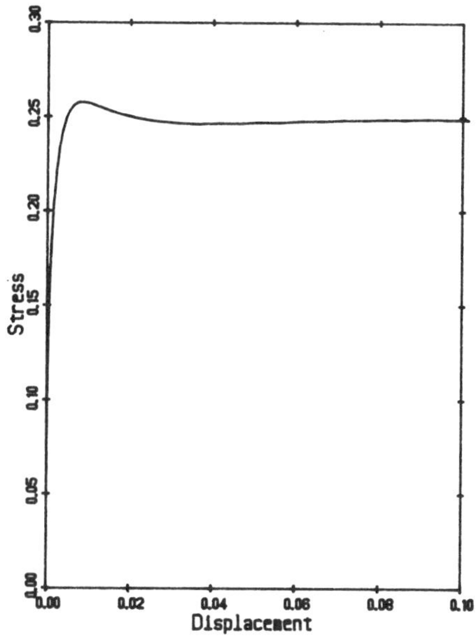

Figure 6: Stress vs. displacement at $x = 1$, $0 \leq t \leq 0.45$ for Example 3.

$PrV_0^2(1-x^2)/2$. If the maximum value of this temperature profile is below θ_m then the slab will shear uniformly. However, if the maximum temperature is above θ_m, a thermally softened layer will occur near the center of the slab.

We illustrate this effect by solving a problem with equation (4.2a) and $Pr^{-1} = 0.1$, $Re^{-1} = 0.1$, $\Delta\theta = 0.01$, $\theta_m = 0.03$, $G_\infty = 0.05$, $V_0 = 0.2$, $r = 0.05$, $d \to \infty$.

The solution (cf. Figure 7) shows significant shear in the center of the slab, especially in the velocity profile. It is interesting to note that the temperature spike is spreading due to conductivity, and that this broadens the region of high shear. The engineering stress - displacement plot (cf. Figure 7d) shows the initial overshoot, followed by an interval of constant stress, followed in turn by damped "stress jumps" due to reflected damped elastic waves arriving at $x = 1$. In the absence of viscosity, these "stress jumps" would be discontinuous. Here, there is sufficient viscosity so that they are smoothed to the point of resembling a linear

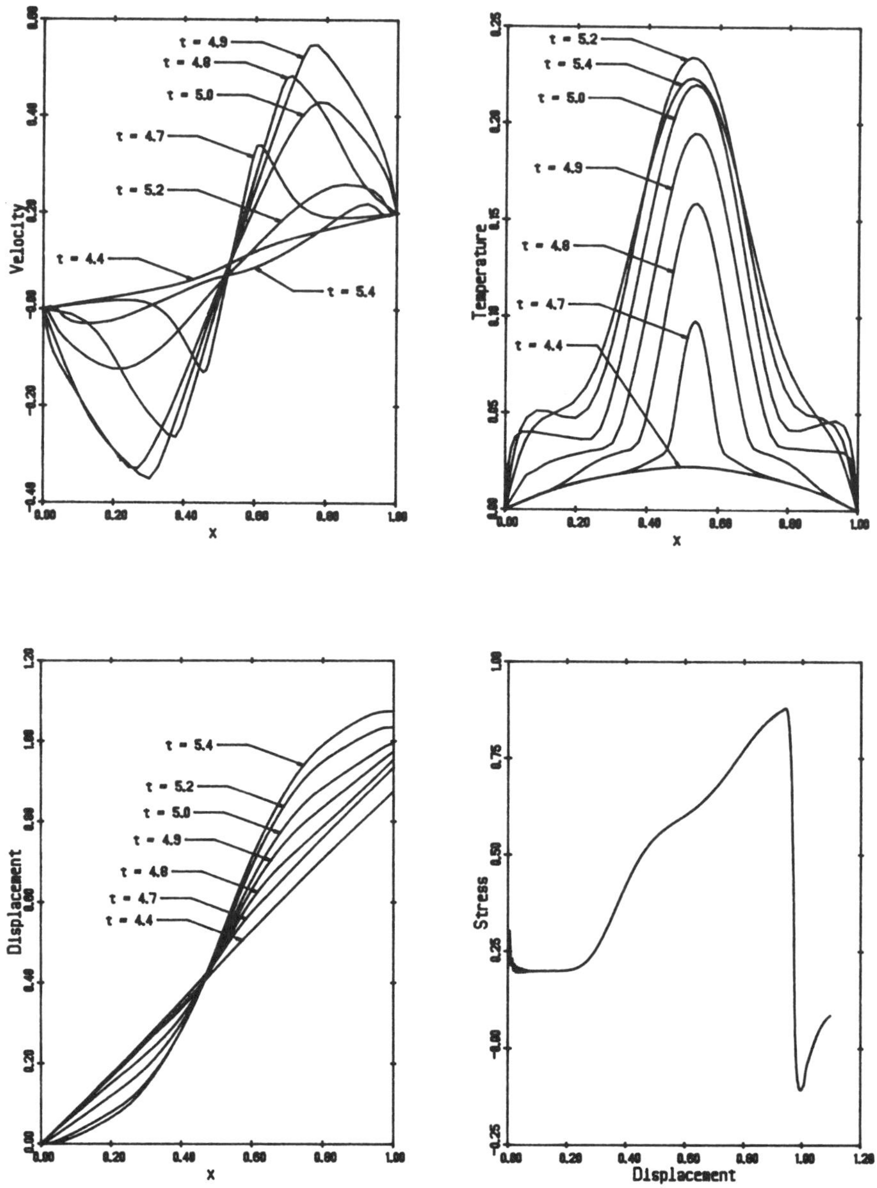

Figure 7: Velocity, temperature, and displacement vs. x at t = 4.4, 4.7, 4.8, 4.9, 5.0, 5.2, and 5.4 and stress vs. displacement at x = 1, $0 \leq t \leq 5.4$ for Example 4.

static stress-strain law. Finally, when the temperature reaches θ_m, the stress drops sharply, indicating that the slab has lost most of its strength.

5. DISCUSSION.

We have presented several examples of the numerical solution of one-dimensional initial-boundary value problems where the formation of locallized regions of high temperature results in thermal softening which, in turn, results in regions of large displacement and velocity gradients, or shear bands. We used a moving grid finite element method to obtain results in four examples that illustrate different aspects of the physical situation. The finite element code performed adequately, and showed that it is capable of resolving rapidly developing large gradients structures without numerical instability. Minor oscillations in the mesh trajectories were confined to regions where the discretization error was insignificant.

We suggest three mechanisms for material failure due to self-induced thermal softening. The first is the immediate "shear off" mode where the conditions are such that the first passage of an elastic wave generates enough heating to cause a shear layer near the sheared end. A second mode of failure is less immediate, involving the heat build-up due to the passage of one or more elastic waves and failure due to an unloading wave and the resultant heat generation. Both these short-term failures involve temperature maxima near the edge of the slab being sheared.

The third failure mode takes a longer time to develop. It involves thermal evolution towards a steady-state profile with maximum temperature occuring near the center. The elastic waves have dissipated, resulting in a linear shear profile which generates heat.

We are looking forward to extending our simplified shear band model in several directions. First, it would be interesting to study the effect of including the complete energy equation (2.11d) instead of the heat conduction equation (2.11d'). Second, plastic behavior is central to understanding shear bands, and we plan to introduce a plastic stress-strain model into our formulation. Hopefully, we will also be able to consider higher dimensional models in the future.

The use of general purpose software to solve partial differential equations is new compared to the state of affairs in ordinary differential equations. It offers several advantages over special purpose software whenever computational speed is not of paramount importance. For example, it is easy to change physical models in order to try new ideas. Furthermore, scientists and engineers need not have a thorough knowledge of sophisticated numerical methods or computing.

We are continuing our research on adaptive software and we feel that we have reached the limit in terms of what can be achieved with a fixed number of elements per time step. We are examining two possible extensions of our methods to include the possibility of adding and deleting elements as the integration progresses in time. The first technique is a method of lines approach, in the spirit of Bieterman and Babuska [3,4]; however, the "lines" adaptively move so as to minimize artificial dispersion effects. Elements can be added or deleted as the integration progresses and the existing codes for ordinary differential equations can still be used. A second technique is based on local refinement, using space-time trapezoidal elements. This is in the spirit of the adaptive finite difference methods of Berger [2]; however, the computational cells are trapezoidal rather than rectangular. Work on this latter approach is well underway and will be reported on shortly [11].

REFERENCES

1. Babuska, I., J. Chandra, and J. E. Flaherty (Eds.), Adaptive Computational Methods for Partial Differential Equations, SIAM, Philadelphia, 1983.
2. Berger, M. J., Adaptive mesh refinement for hyperbolic partial differential equations, Report No. STAN-CS-82-924, 924, Department of Computer Science, Stanford University, 1982.

3. Bieterman, M. and I. Babuska, The finite element method for parabolic equations, I. a posteriori error estimation, Numer. Math., 40 (1982), pp. 339-371.

4. M. Bieterman and I. Babuska, "The finite element method for parabolic equations, II. a posteriori error estimation and adaptive approach", Numer. Math., 40 (1982), pp. 373-406.

5. Costin, L. S., E. E. Crisman, R. H. Hawley, and J. Duffy, On the localization of plastic flow in mild steel tubes under dynamic torsion, Inst. Phys. Conf. Ser. No. 47: Chapter 1, Inst. of Physics, 1979, pp. 90-100.

6. Coyle, J. M., J. E. Flaherty, and R. Ludwig, On the stability of mesh equidistribution strategies for time dependent partial differential equations, in preparation.

7. S. F. Davis, An adaptive grid finite element method for initial-boundary value problems, Ph.D. Dissertation, Rensselaer Polytechnic Institute, 1980.

8. Davis, S. F. and J. E. Flaherty, An adaptive finite element method for initial-boundary value problems, SIAM J. Sci. and Stat. Comp., 3 (1982), pp. 6-27.

9. de Boor, C., A Practical Guide to Splines, Applied Mathematical Sciences, No. 27, Springer-Verlag, New York, 1978.

10. Flaherty, J. E., J. M. Coyle, R. Ludwig, and S. F. Davis, Adaptive finite element methods for parabolic partial differential equations, in I. Babuska, J. Chandra, and J. E. Flaherty (Eds.), Adaptive Computational Methods for Partial Differential Equations, SIAM, Philadelphia, 1983.

11. Flaherty, J. E., and P. K. Moore, An adaptive local refinement finite element method for parabolic partial differential equations, to appear in Proc. Int. Conf. on Accuracy Estimates and Adaptive Refinements in Finite Element Computations, Lisbon, Portugal, 1984.

12. Hedstrom, G. W., and G. H. Rodrique, Adaptive-grid methods for time-dependent partial differential equations, UCRL87242 preprint, Lawrence Livermore National Laboratory, Livermore, 1982.

13. Jamet, P. and R. Bonnerot, Numerical solution of the Eulerian equations of compressible flow by a finite element method which follows the full boundary and the interfaces, J. Comp. Phys. 18 (1975), pp. 21-45.
14. Pereyra V. and E. G. Sewell, Mesh selection for discrete solution of boundary problems in ordinary differential equations, Numer. Math., 23 (1975), pp. 261-268.
15. L. Petzold, Personal communication.
16. Wheeler, M. F., A priori L_2-error estimates for Galerkin approximations to parabolic partial differential equations, SIAM J. Numer. Anal., 10 (1973), pp. 723-759.

This work was partially supported by the U. S. Army Research Office under contracts DAAG28-82-K-0185 and DAAG29-82-K-0197 and the Air Force Office of Scientific Research, Air Force Systems Command, under grant number AFOSR-80-0192. The authors would like to thank Ms. Diane McNulty for the fine job of typing and her good humor.

Department of Mathematical Sciences
Rensselaer Polytechnic Institute
Troy, New York 12181

THE CAUCHY AND BORN HYPOTHESES FOR CRYSTALS

J. L. Ericksen

1. INTRODUCTION.

Commonly, molecular theories of crystal elasticity lean upon hypotheses introduced by Cauchy [1-3] or Born [4] to relate changes in atomic positions to macroscopic deformation. Both men pictured the atoms as mass points. Briefly and roughly, Cauchy assumed that atomic motion and gross motion are the same, where both are defined. Later, it was appreciated, in particular by Born, that, in a solid which appears to be at rest, atoms still undergo vibratory (thermal) motions about equilibrium positions. Since such things as x-ray observations average out such fluctuations, they can appear to be in good agreement with Cauchy's hypothesis. By these standards, Cauchy's hypothesis might or might not describe deformations encountered in transitions observed in crystals. For purposes of discussion, I will ignore such fluctuations. Born pointed out that, in some cases, Cauchy's hypothesis is in trouble for a different reason, being inconsistent with certain conditions of equilibrium. As an alternative, he proposed that lattice vectors deform as would material line elements, subject to the macroscopic deformation, the aforementioned equilibrium conditions being used to fix the finer details of atomic arrangement. In particular cases, this leads to deformations consistent with Cauchy's hypothesis. Commonly, studies requiring such an hypothesis use one of the two.

ISBN 0-12-309770-3

In trying to apply, or to decide whether either hypothesis is applicable to deformations involved in phase transformations, one encounters ambiguities, complicating the matter. My primary purpose is to elaborate this.

2. THE HYPOTHESES.

A classical definition of crystals pictures configurations of atoms, filling all of space. To be crystal configurations, these must have a periodic structure, described by a translation group, generated by three (constant) linearly independent vectors, a_1, a_2 and a_3. The idea is that any point must be carried to a physically indistinguishable point by all translations of the form

$$n^K a_K \ , \tag{2.1}$$

where we use the summation convention, and the n^K represent any set of integers. What is sometimes left unsaid, but to be understood, is that this group is maximal; we don't skip over any indistinguishable points. For example, we commonly picture amorphous solids as homogeneous, meaning that we have such a translation group for any choice of a_K, but a crystallographer would not include them. With this understanding, the a_K are called lattice vectors. Less than maximal groups are sometimes encountered in practice, as will become clear. Different sets of lattice vectors are equivalent, in the sense that they generate the same translation group. For two sets a_K and $\bar{a}_K$ to be equivalent, it is necessary and sufficient that they be related by an equation of the form

$$\bar{a}_K = m_K^L a_L \ , \tag{2.2}$$

where the m_K^L are any integers such that

$$\det.\| m_K^L \| = \pm 1 \ , \tag{2.3}$$

describing the equivalence as a representation of an infinite discrete group G, which plays an important role in the classical theory of crystallographic groups. A monatomic crystal may or may not occur as a so-called simple or Bravais lattice, doing so provided application of the translation group to one atom gives the positions of all.

Of course, we observe bodies which are not infinite, but, finite. If a body, or some macroscopic part of it, can reasonably be identified with a restriction of the ideal infinite crystal, the part passes as a crystal. In the process, one exercises some judgment about defects occurring in real crystals. One crystal might, after a phase transition, exist as a number of crystals somehow joined together. It can be hard to know just what part of the original configuration corresponds to a given part of the final, let alone decide exactly what deformation it experienced. Clearly, deciding which of an identical set of atoms goes where involves some guesswork. As is discussed by Nishiyama [5, Ch. 6], for example, metallurgists have had some limited success unravelling such puzzles. Among the ideas used is Cauchy's. Actually, some bodies of interest consist of parts which are not true crystals, but strongly resemble them. For, say, common carbon steels, the iron atoms can form a good simple lattice, but carbon atoms are distributed rather randomly. For present purposes, such things might well be regarded as crystals.

In dealing with the ideal infinite configurations, it seems fairly natural to assume, as Born did, that deformations taking one to another are homogeneous, the deformation gradient F being constant. Here we assume that some such configuration is taken as a reference, with some definite choice of reference lattice vectors A_K. The Born hypothesis then reads

$$a_K = FA_K \ , \qquad (2.4)$$

the a_K being a possible set of lattice vectors in the deformed crystal. Given A_K and F, we clearly get just one of the infinitely many possible choices of lattice vectors for the deformed crystal, it being a matter of chance whether these are the same lattice vectors which an x-ray crystallographer would select. Clearly, (2.4) describes a linear transformation so, in particular, for any element of G, we have

$$m_K^L a_K = F(m_K^L A_K) \ , \qquad (2.5)$$

from which one can see that the validity of the Born hypothesis does not really depend upon a special choice of the reference lattice vectors. For a simple lattice deforming to a simple lattice, (2.4) rather suggests, as many would assume, that the atom originally at $n^K A_K$ moves to $n^K a_K$, ignoring a trivial translation. If so, it is consistent with Cauchy's hypothesis. Conversely, it is not hard to show that an homogeneous deformation taking a simple lattice to another will, if it is consistent with Cauchy's hypothesis, satisfy (2.4), for some choice of lattice vectors. Similar agreement occurs in some other cases, not in others. Of course, in itself, (2.4) says nothing about the fate of individual atoms, only about the periodicity of sets which they form. Thus, some assumptions are added, in making such comparison.

It seems pretty clear that (2.4) fails to apply to some of the kinds of continuous or second-order transformations considered by Landau [6], who did not discuss F or the equivalent, avoiding the need for accepting any particular hypothesis about it. Briefly and roughly, atomic positions are assumed to shift in a continuous way with pressure and temperature, but there can be, say, a sudden doubling in length of a lattice vector. More precisely, no matter how we select lattice vectors, at least one experiences a sizeable discontinuity. To see that this is possible one need only appreciate that the precise periodicity can be changed considerably, by infinitesimal shifts in positions of some atoms. Here, (2.4) would require F to suffer an unbelievably large discontinuity. Cauchy's hypothesis seems much more reasonable, on the face of it. To accept it, one must argue that the macroscopic F can vary somewhat over distances of the order of a few atomic spacings, which induces some queasiness.

The latter type of difficulty becomes still more severe in so-called shuffle transformations. Here, lattice vectors, chosen in an obvious way, remain fixed. So does F, as observations are interpreted. This is consistent with (2.4), so Born's hypothesis can be considered to apply. However, some atoms in a unit cell undergo finite displacements, to symmetry-related positions. Certainly, it does not seem very reasonable to consider that Cauchy's hypothesis is applicable to such cases.

While these hypothesis have their faults, they deserve serious consideration, so we should clearly understand just what they imply. In part, this is complicated by ambiguities inherent in either. We now focus on some associated with the Born hypothesis, there being rather similar kinds associated with Cauchy's.

3. LATTICE-INVARIANT DEFORMATIONS.

Even when the Born rule applies, with F constant, the previous discussion makes clear that measurements of lattice vectors alone do not suffice to determine F uniquely. Said differently, infinitely many homogeneous deformations, which I call lattice-invariant deformations, take the ideal infinite crystal onto itself, in a manner consistent with the Born rule.

Consider any crystal configuration as a reference, and take any possible set of lattice vectors a_K as a reference set. From (2.2) and (2.4), we see that F can reasonably be considered to describe a lattice-invariant deformation provided there is some element of the group G such that

$$Fa_K = m_K^L a_L \ . \tag{3.1}$$

Mathematically, such F merely form a different representation of G or, if you like, a conjugate group. Introducing the dual basis a^K, the so-called reciprocal lattice vectors, such that

$$a_K \cdot a^L = \delta_K^L \ , \qquad a_K \otimes a^K = 1 \ , \tag{3.2}$$

we can solve (3.1) for F, obtaining

$$F = m_K^L a_L \otimes a^K \ . \tag{3.3}$$

Commonly, F is understood to be orientation preserving, so

$$\det F > 0 \ , \tag{3.4}$$

Assuming this, we are restricted to the subgroup of G

whose elements have positive determinant and, using (2.3), we find that (3.3) implies that

$$\det F = 1 . \tag{3.5}$$

Since such deformations take the infinite crystal to an indistinguishable configuration, superposing a lattice-invariant deformation on any deformation should leave invariant such things as elastic strain energy functions or associated Cauchy stress tensors, at least as I and some others see it. Molecular theory of elasticity seems to support this view. When Nishiyama [5, p. 339] argues that the Bain deformation is most reasonable because it has the smallest strain energy, he seems to espouse a contrary view. Given this and other similar statements in the metallurgical literature, it seems unfair to claim that the assertion is commonly accepted. Parry [7-8] and Pitteri [9] present analyses helpful in constructing constitutive equations exhibiting the aforementioned invariance. Of course, Born's hypothesis might hold in cases where elasticity theory fails to apply. Rivlin [10] discusses some cases which might be regarded as illustrating the possibility. Then, the usual ideas of strain energy and stress require some modification, to fit some different kind of theory.

One can push the motion a bit further, to consider the possibility that F is not constant, but piecewise constant. This seems reasonable, as long as the diameter of a set on which F is constant is reasonably large, compared to atomic spacing. Roughly, this is measured by the lengths of lattice vectors, selected to be a short as possible. In a rather natural way, this leads to a notion of lattice-invariant shears, similar to that used by metallurgists, in attempts to describe rather complex deformations encountered in some martensitic transformations. The discussion by James [11] is likely to be more accessible to those trained in mathematics or continuum mechanics.

Suppose that two neighboring parts undergo homogeneous deformations relative to some homogeneous configuration, with deformation gradients F_1 and F_2. We can define the relative deformation gradient F by

$$F_2 = FF_1 , \tag{3.6}$$

what would be the deformation gradient if we took as a reference the obvious homogeneous extrapolation of the first part. Picture some part of this reference as undergoing the deformation corresponding to F, with the gradient having a finite discontinuity on a plane with unit normal initially ν, say. We assume that the displacement remains continuous. The usual kinematical conditions of compatibility then imply that

$$F = 1 + \alpha \otimes \nu , \tag{3.7}$$

where α is some constant vector, not the null vector. Actually, the same condition obtains if we assume that the displacement has a constant jump discontinuity, as might be associated with slip, one of the possibilities considered by metallurgists. Second, we assume that F is a restriction of a lattice invariant deformation. This is, if a_K are lattice vectors in the first region, we must have, for some element of G,

$$F = 1 + \alpha \otimes \nu = m_K^L a_L \otimes a^K . \tag{3.8}$$

Here, a definite choice of a_K might be dictated to be obtained from the original reference set by applying F, using the Born rule. As an x-ray crystallographer would see it, lattice vectors then remain continuous.

In particular, (3.5) now applies, giving

$$\det F = 1 \iff \alpha \cdot \nu = 0 . \tag{3.9}$$

Thus, F has the form commonly associated with a simple shearing deformation, making it natural to call these lattice-invariant shears. Some metallurgists seem to use the term to include such things as martensitic twinning, involving a discontinuity in lattice vectors which is quite apparent from x-ray observations. According to theories of elasticity, invariant in the manner indicated, it is not automatic that stresses in such twins will be the same, although one will be unstressed if the other is. Common analyses of these employ a relation which is similar to, but different from (3.8). Some special features of equilibrium under zero stress are discussed by James [11]. Similar considerations apply to cases where the Cauchy stress reduces to an hydrostatic pressure.

Particular examples can be defined by

$$Fa_1 = a_1, \quad Fa_2 = na_1 + a_2, \quad Fa_3 = a_3 , \tag{3.10}$$

where n is any integer, a_K any linearly independent vectors, considered as lattice vectors. Here, solving for F gives

$$F = 1 + na_1 \otimes a^2 . \tag{3.11}$$

Here, we have

$$\begin{aligned} \alpha &= \pm na_1 \| a^2 \| , \\ \nu &= \pm a^2 / \| a^2 \| . \end{aligned} \tag{3.12}$$

Clearly, we here take

$$\| m_K^L \| = \begin{Vmatrix} 1 & 0 & 0 \\ n & 1 & 0 \\ 0 & 0 & 1 \end{Vmatrix} , \tag{3.13}$$

which, being a unimodular matrix of integers, is in G. In a certain sense, all possibilities are of this kind. Suppose we have F, a_K, and m_L^K, satisfying (3.1). We can take any other element of G, say $\bar{m}_L^K$, and define an equivalent set of lattice vectors $\bar{a}_K$ by

$$\bar{a}_K = \bar{m}_K^L a_L , \tag{3.14}$$

with

$$\begin{aligned} F\bar{a}_K &= \bar{m}_K^L F a_L = \bar{m}_K^L m_L^P a_P , \\ &= \bar{m}_K^L m_L^P (\bar{m}^{-1})_P^Q \bar{a}_Q , \\ &= \bar{\bar{m}}_K^L \bar{a}_L . \end{aligned} \tag{3.15}$$

Here, we have denoted in an obvious way the inverse of the group element used. Clearly, $\bar{\bar{m}}_K^L$ represents an element of G, obtained from m_K^L by applying a similarity transformation, selected as an arbitrary element of G. The allegation is that, given any solution to the problem indicated, we can use such similarity transformations to reduce it to the form described by (3.10)-(3.10).

To begin to establish this, solve (3.8) for m^K_L, which gives

$$m^L_K = \delta^L_K + \alpha^L \nu_K \ , \tag{3.16}$$

where

$$\left.\begin{aligned} \nu_K &= \nu \cdot a_K \ , \\ \alpha^L &= \alpha \cdot a^L \ . \end{aligned}\right\} \tag{3.17}$$

We have, using (3.2) and (3.9),

$$\alpha^L \nu_L = \alpha \cdot a^L \otimes a_L \nu = \alpha \cdot \nu = 0 \ . \tag{3.18}$$

Generally, α^L and ν_K won't be integers, but the m^L_K must be. For any $\lambda \neq 0$, we will have

$$\beta^L \mu_K = \alpha^L \nu_K \tag{3.19}$$

if

$$\beta^L = \lambda \alpha^L , \quad \mu_K = (1/\lambda) \nu_K \ . \tag{3.20}$$

By properly choosing λ, we can arrange that the β^L are all integers, the μ_K rational numbers. For example, if $\nu_1 \neq 0$, take $\lambda = 1/\nu_1$. In particular, we then know that the quantities

$$\alpha^L \nu_1 = \beta^L \mu_1 = \beta^L \tag{3.21}$$

must be integers. Assuming this, if, say, $\beta^2 \neq 0$, we know that

$$\beta^2 \ \mu_K \tag{3.22}$$

must be integers, requiring the μ_K to be rational numbers. One can say more about them, but this will suffice. Similarity transformations of the type allowed will take

$$m^L_K = \delta^L_K + \beta^L \mu_K \tag{3.23}$$

to elements of the same form, say

$$\bar{\bar{m}}^L_K = \delta^L_K + \bar{\bar{\beta}}^L \bar{\bar{\mu}}_K \tag{3.24}$$

where the $\bar{\bar{\beta}}^L$ are linear functions of the β^K with integer coefficients, $\bar{\bar{\mu}}_K$ similarly related to μ_K. Thus the former will again be integers, the latter rationals. Also, (3.18) implies that

$$\beta^K \mu_K = \bar{\bar{\beta}}^K \bar{\bar{\mu}}_K = 0 . \tag{3.25}$$

Starting with this information, it is straightforward to use an elementary theorem in number theory to construct algorithms for calculating the similarity transformations needed to effect the indicated reduction, for the various possible cases.

CASE 1: Two of the β^K vanish.

By a possible transformation, renumbering lattice vectors, we can assume that

$$\beta^1 \neq 0, \ \beta^2 = \beta^3 = 0 , \tag{3.26}$$

whence follows from (3.25) that

$$\mu_1 = 0 . \tag{3.27}$$

If either μ_2 or μ_3 vanishes again renumber to get $\mu_3 = 0$. We then know that

$$\beta^1 \mu_2 = n ,$$

where n is some integer. This gives us (3.13). If $\mu_2 \mu_3 \neq 0$, we know that, for some integers n_2 and n_3, we have

$$\begin{aligned} \beta^1 \mu_2 &= n_2 , \\ \beta^1 \mu_3 &= n_3 , \end{aligned} \tag{3.28}$$

If n is the greatest common divisor of these integers, so that the integers

$$p_2 = n_2/n, \quad p_3 = n_3/n \tag{3.29}$$

are relatively prime, it is an elementary theorem in number theory that there exist integers q and r such that

$$p_2 q - p_3 r = 1 . \tag{3.30}$$

A possible transformation, described in terms of reciprocal lattice vectors, is then given by

$$\left.\begin{aligned} \bar{a}^1 &= a^1 , \\ \bar{a}^2 &= p_2 a^2 + p_3 a^3 , \\ \bar{a}^3 &= r a^2 + q a^3 , \end{aligned}\right\}$$

which implies that $\bar{a}_1 = a_1$. Thus,

$$\left.\begin{aligned} F &= 1 + \beta^1 a_1 \otimes (\mu_2 a^2 + \mu_3 a^3) , \\ &= 1 + n\bar{a}_1 \otimes (p_2 a^2 + p_3 a^3) , \\ &= 1 + n\bar{a}_1 \otimes \bar{a}^2 , \end{aligned}\right\} \tag{3.31}$$

giving us the reduction to (3.11).

<u>CASE 2:</u> <u>One of the</u> β^K <u>vanishes</u>.

As before, renumber to get $\beta^3 = 0$. With m as the greatest common divisor of β^1 and β^2, write

$$\beta^1 = mp^1, \quad \beta^2 = mp^2 , \tag{3.32}$$

now letting q and r be integers such that

$$p^1 q - p^2 r = 1 . \tag{3.33}$$

With the allowable change of lattice vectors given by

$$\left.\begin{aligned} \bar{a}_1 &= p^1 a_1 + p^2 a_2 , \\ \bar{a}_2 &= r a_1 + q a_2 , \\ \bar{a}_3 &= a_3 , \end{aligned}\right\} \tag{3.34}$$

we have

$$\begin{aligned} \beta^K a_K &= m(p^1 a_1 + p^2 a_2) = m\bar{a}_1 \\ &= \bar{\beta}^K \bar{a}_K , \end{aligned} \tag{3.35}$$

or $\bar{\beta}^2 = \bar{\beta}^3 = 0$. Applying to this the analysis described in CASE I then gives the desired reduction.

<u>CASE 3:</u> <u>The</u> β^K <u>are all non-zero</u>.

Here write

$$\beta^2 = mp^2, \quad \beta^3 = mp^3 , \tag{3.36}$$

with p^2 and p^3 relatively prime, now choosing integers q and r so that

$$p^2 q - p^3 r = 1 . \tag{3.37}$$

The change of lattice vectors given by

$$\left.\begin{aligned} \bar{a}_1 &= a_1 , \\ \bar{a}_2 &= p^2 a_2 + p^3 a_3 , \\ \bar{a}_3 &= r a_2 + q a_3 , \end{aligned}\right\} \tag{3.38}$$

then gives

$$\begin{aligned} \beta^K a_K &= \beta^1 a_1 + m(p^2 a_2 + p^3 a_3) \\ &= \beta^1 \bar{a}_1 + m\bar{a}_2 = \bar{\beta}^K \bar{a}_K , \end{aligned} \tag{3.39}$$

reducing this to CASE 2.

To sum up, if β^K and μ_L are, respectively, integers and rational numbers, (3.23) defines an element of G provided the numbers also satisfy (3.25) and the condition that the products $\beta^K \mu_L$ be integers. With a_K taken as a possible selection of lattice vectors, F, given by (3.8) defines one of the possible lattice-invariant shears. By using the above algorithms, we can always find lattice vectors reducing any possible F to the form (3.11). Said differently, ν must be parallel to a possible reciprocal lattice vector, α to one of the corresponding perpendicular lattice vectors, its magnitude being limited by (3.12).

Rather obviously, such discontinuities give rise to discontinuities in lattice vectors which agree with the Born rule, although the x-ray crystallographer would perceive lattice vectors as constant throughout. Perhaps it only reflects my lack of ingenuity, but I find it difficult to seriously consider anything more general than piecewise homogeneous deformations, as a real ambiguity involved in relating the Born hypothesis to measurements of lattice vectors. Of course, the metallurgist uses other clues, pondering how finite crystals are seen to be shaped, fit together, etc. It is not entirely easy to sort out all of the mathematical and mechanical ideas which might be involved in such considerations.

4. PRACTICE.

It is not so hard for the uninitiated to be misled by common descriptions of configurations. In a certain temperature range, a monatomic crystal might adopt a configuration sometimes described as a face-centered cubic. At other temperatures, it might occur in the form described as body-centered cubic. Different phases of iron are of these forms, for example. The words suggest picturing an homogeneous deformation of one cube to another. Apart from a rather inconsequential translation and rotation, this could only be a uniform dilation, all directions being stretched the same, a conformal mapping. Experience contradicts this, and it is inconsistent with the Cauchy and Born hypotheses.

For the face-centered cubic, the words suggest a translation group generated by three orthogonal vectors of equal length; forming edges of a cube, say b_1, b_2 and b_3. Application of this translation group to one atom, located at the origin, would generate a simple lattice, a simple cubic configuration. To get the face centered variety, we add identical atoms at three places, say

$$\tfrac{1}{2}(b_1 + b_2),\ \tfrac{1}{2}(b_2 + b_3),\ 1/2(b_3 + b_1) \quad , \tag{4.1}$$

applying the same translation group to these, to locate positions of remainder. Or, equivalently, we place atoms at positions whose components, relative to this basis, are either integers or half-integers.

In discussing deformations likely to occur in transitions of the kind mentioned, metallurgists sometimes picture the configuration in a different way, as a body-centered tetragonal. Here one introduces as translation vectors c_K, given by

$$\left.\begin{aligned} c_1 &= \tfrac{1}{2}(b_1 - b_2) \ , \\ c_2 &= \tfrac{1}{2}(b_1 + b_2) \ , \\ c_3 &= b_3 \ , \end{aligned}\right\} \tag{4.2}$$

still orthogonal, with c_1 and c_2, but not c_3 of equal length, describing edges of the tetragon. With the factors of $1/2$ occurring, the two sets are not related by G. From this alone, it follows that not both can be lattice vectors. Application of the second group to an atom at $1/2 (b_1 + b_2)$ generates most atomic positions. To get the rest, similarly translate an atom whose position vector, relative to this point, is

$$1/2 (b_1 + b_2 + b_3) . \tag{4.3}$$

Another way of describing the same configuration is to introduce translation vectors

$$\left.\begin{aligned} a_1 &= 1/2 (b_1 - b_2) = c_1 , \\ a_2 &= 1/2 (b_1 + b_2) = c_2 , \\ a_3 &= 1/2 (b_1 + b_3) = 1/2 (c_1 + c_2 + c_3) , \end{aligned}\right\} \tag{4.4}$$

Applied to one atom, this generates the whole set, describing it as a simple lattice, these a_K being one of the possible sets of lattice vectors. With the factors of $1/2$ running in (4.4), neither b_K nor c_K can be lattice vectors. In the jargon used by Ericksen [12], they are sub-lattice vectors. Generally sub-lattice vectors b_K are related to lattice vectors by equations of the form

$$b_K = n_K^L a_L , \tag{4.5}$$

where the n_K^L are integers such that

$$\det \| n_K^L \| \neq 0, 1, -1 , \tag{4.6}$$

so the inverse transformation exists, with coefficients which are rational numbers, not all integers. In at least some cases where the Born hypothesis fails, a modification can reasonably be applied, with lattice vectors replaced by suitably selected sub-lattice vectors. Of course, this makes the hypothesis still more ambiguous.

For the body-centered cubic, we also introduce three orthogonal vectors d_K of equal length, like the previous b_K. From an atom at the origin, this again generates a simple cubic lattice. Add an atom at

$$1/2 (d_1 + d_2 + d_3) , \tag{4.7}$$

and similarly translate it to complete the configuration. Again, this is a simple lattice, with one set of lattice vectors $\hat{a}_K$ being given by

$$\left.\begin{aligned} \hat{a}_1 &= d_1 , \\ \hat{a}_2 &= d_2 , \\ \hat{a}_3 &= \tfrac{1}{2}(d_1 + d_2 + d_3) , \end{aligned}\right\} \tag{4.8}$$

and, again, the d_K are sub-lattice vectors, but not lattice vectors. The descriptions as face- or body-centered cubes have some merit, making rather obvious the crystallographic point groups appropriate for these. By a routine calculation, one can get this from the simple lattice description. The body-centered tetragonal description suggests a different point group, and it is less routine to correctly calculate the point group, using it, but it makes it easier to picture some relevant deformations. According to either the Cauchy or the Born hypothesis, which here agree, a possible homogeneous deformation taking the face-centered cubic to the body-centered cubic configuration is defined in terms of lattice vectors described above by, as the deformation with gradient F such that

$$\hat{a}_K = Fa_K . \tag{4.9}$$

Of course, one can superpose lattice-invariant deformations, as described earlier, including use of lattice-invariant shears. On theoretical grounds, I see no easy way of deciding that one of these is more likely to be observed than another, although one expects them to be separated by energy barriers. With (4.9), elementary calculations indicate that one can picture the deformation as that taking the tetragon with edges c_K to the cube with edges d_K, a deformation which is clearly different from the uniform dilatation mentioned at the beginning of this section. In the metallurgical literature, this is called the Bain distortion or Bain deformation. Some discussions of this such as that of Nishiyama [5, p. 339] mention experimental confirmation that this is the deformation which occurs in at least some cases. Following this is a discussion of martensitic transformations which involve much more complex patterns of deformation, with quite different crystal configurations contacting each other. Certainly, it would be nice to have better tools, to resolve such puzzles.

In this discussion, I glossed a point. Conceivably, the uniform dilatation and the deformation given by (4.9) could both be consistent with the Born hypothesis. It is not very hard to show that they can't, but I won't take space to elaborate this.

REFERENCES

1. Cauchy, A.-L., Sur l'equilibre et le mouvement d'un système de points materiels sollicités par forces d'attraction ou de répulsion mutuelle, Ex. de Math 3 (1828), 227-287.
2. Cauchy, A.-L., De la pression on tension dans un système de points matériels, Ex. de Math. 3 [1828), 253-277.
3. Cauchy, A.-L. Sur les équations differentielles d'equilibre ou de mouvement pour un système de points matériels, sollicités par des forces d'attraction ou de répulsion mutuelle, Ex. de Math. 4 (1829), 342-369.
4. M. Born, Dynamik der Kristallgitter, B. G. Teubner, Leipzig-Berlin (1915).
5. Z. Nishiyama, Martensitic Transformation, Academic Press, New York-San Francisco-London (1978).
6. L. D. Landau, On the theory of phase transitions, in Collected Papers of L. D. Landau (ed. D. Ter Haar) Gordon and Breach and Pergamon Press, New York-London-Paris (1965).
7. Parry, G. P., On the elasticity of monatomic crystals, Mat. Proc. Camb. Phil. Soc. 80 (1976), 189-211.
8. Parry, G. P., On diatomic crystals, Int. J. Solids Structures 14 (1978), 283-287.
9. Pitteri, M. Reconciliation of local and global symmetries of crystals, to appear in J. Elasticity.

10. Rivlin, R. S., Some thoughts on material stability, in Finite Elasticity (D. E. Carlson and R. T. Shield, eds.), Martinus Nijhoff Publishers, Hague-Boston-London-London (1970), 105-122.
11. James, R. D., Mechanics of coherent phase transformations in solids, Materials Research Laboratory Report MRL E-143, Brown University, Providence, October 1982.
12. Ericksen, J. L., Crystal lattices and sub-lattices, Rend. Sem. Mat. Univ. Padova, 68 (1982), 1-9.

Department of Aerospace Engineering and Mechanics
and School of Mathematics
University of Minnesota
Minneapolis, MN 55455

The Arrangement of Coherent Phases in a Loaded Body

R. D. James

1. INTRODUCTION.

If the temperature of certain crystals is lowered, tiny platelets and needles of a new phase grow into the crystal from its boundary. Each platelet is separated from the parent phase by a surface which is often idealized as a surface of discontinuity of the strain, but no diffusion takes place. These transformations are described as martensitic in the materials science literature and as polymorphic or displacive in the geological literature.

It is the purpose of this paper to propose a theory for the response of materials which can undergo such transformations, and to explore its simplest consequences for dead loaded bodies. In §3 I describe some observations of a material which undergoes a martensitic transformation. At the transformation temperature, 24 variants of a new phase appear, each adjoining the parent phase. These are arranged in a manner consistent with certain kinematic rules described in §1. If a uniaxial tension is applied, only one of the 24 variants survives; it grows at the expense of the other variants and consumes the whole specimen at quite small values of the applied load. These observations suggest a form of the constitutive equation which I describe in §4. A result discussed in §1 gives simple necessary and sufficient conditions for certain piecewise linear deformations to be stable in a dead loading device; I am

ISBN 0-12-309770-3

able to obtain these conditions only if the arrangement of the phases is not too complicated. This result is the basis of a partial analysis (§6) of whether the parent phase can be recovered by applying dead loads to a body which has already transformed. A full resolution of this and related questions appears to depend on the relationship of two vectors $\underset{\sim}{a}$ and $\underset{\sim}{n}$ used to define the transformation strain, on the symmetry of the parent phase, on the elastic moduli of the parent phase, and on the elastic moduli of one variant of the daughter phase. The method of attack involves convex analysis on certain nine dimensional sets.

2. SOME GEOMETRIC AND MECHANICAL RESTRICTIONS ON THE ARRANGEMENT OF SOLID PHASES

I shall draw on some results of reference [1] in the following sections, so I summarize those results here. The first results are concerned with geometric restrictions on the arrangement of coherent phases. The phase transformation shall be described by a deformation $\underset{\sim}{y} = \underset{\sim}{\chi}(\underset{\sim}{x})$, $\underset{\sim}{x} \in R$, whose gradient is discontinuous across the phase boundaries. Under ordinary conditions, a scratch etched on the surface of an untransformed body will bend sharply at the phase boundary in the transformed body, but it will not break apart. For example, a scratch line ℓ might experience the deformation shown in Fig. 1; the dashed lines are the phase boundaries.

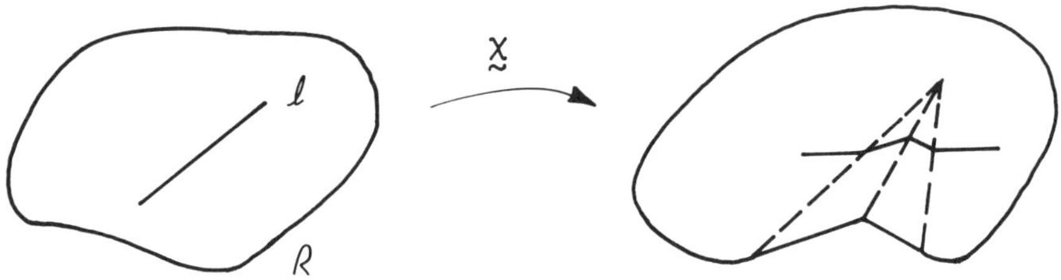

Fig. 1. Deformation of a material line in a transforming body.

Thus, it is reasonable to assume that $\underset{\sim}{X}$ is a continuous function on R whose gradient is continuous everywhere except on surfaces whose images under $\underset{\sim}{X}$ represent the phase boundaries. We also assume that $\nabla\underset{\sim}{X}$ exists as each of these surfaces is approached from either side. Deformations having these properties are termed <u>coherent</u>. If $\underset{\sim}{F}_1$ and $\underset{\sim}{F}_2$ denote limiting values of $\nabla\underset{\sim}{X}$ as $\underset{\sim}{x} \in S$ is approached from each side, a theorem of kinematics guarantees the existence of a vector $\underset{\sim}{a}$ (the <u>amplitude</u>) such that

$$\underset{\sim}{F}_2 - \underset{\sim}{F}_1 = \underset{\sim}{a} \otimes \underset{\sim}{n}. \tag{2.1}$$

In (2.1) $\underset{\sim}{n}$ is a unit normal to S at $\underset{\sim}{x}$. Notice that it is really $\underset{\sim}{X}(S)$ which is identified with the observed phase boundary.

Often, several intersecting phase boundaries are seen in a transformed body. Suppose three phase boundaries meet along a curve, so that the surfaces in R which are deformed into the phase boundaries appear as in Fig. 2.

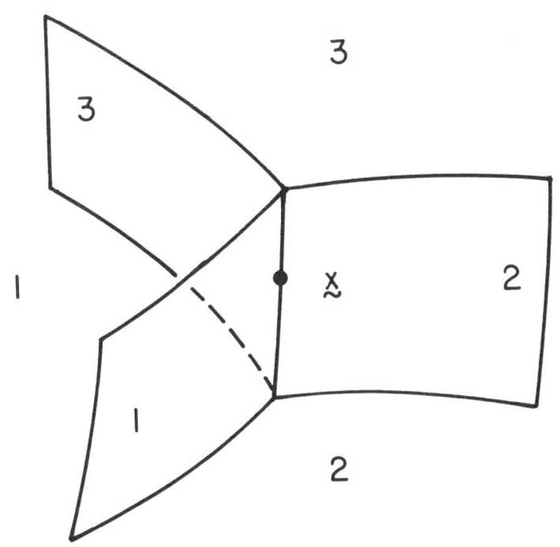

<u>Fig</u>. <u>2</u>. Three phases meeting at a curve.

Let $\underset{\sim}{F}_1$, $\underset{\sim}{F}_2$ and $\underset{\sim}{F}_3$ denote the three deformation gradients obtained by taking the limit of $\nabla\underset{\sim}{X}$ as $\underset{\sim}{x}$ is approached from within each of the regions 1, 2 and 3. Let the surfaces and regions be numbered as in Fig. 2. If we write down equations like (2.1) for the three interfaces and solve them for the amplitudes, we can show that there is a vector $\underset{\sim}{a}$

such that

$$\underset{\sim}{a}_i = \lambda_i \underset{\sim}{a} \qquad i = 1,2,3, \tag{2.2}$$

the λ_i being coefficients in the equation

$$\lambda_i \underset{\sim}{n}_i = 0, \tag{2.3}$$

which follows from the linear dependence of the normals implied by Fig. 2. Notice that if the three deformation gradients are distinct, then none of the λ_i vanish. In this case, according to equation (2.3), no two of the three surfaces have parallel tangent planes at $\underset{\sim}{x}$. It is not uncommon for other considerations like crystallography to imply that the deformation gradients are distinct. In that case, geometric conclusions drawn from an analysis of coherence like the one described above can lead to an understanding of observed arrangements of the phases. For example, it is not possible to have two regions separated by a surface with a corner, subject to a coherent deformation with distinct limiting deformation gradients. A cusp is permissible. This agrees pretty well with observations of transformations truly considered coherent; a single new phase occupying a wedge-shaped region is rare. A desk-top experiment suggests analagous results for more complicated arrangements. If a piece of paper which creases well is crumpled up, flattened with a book and then unfolded, it will contain a variety of two-dimensional coherent arrangements, but no corners.

The arrangement shown in Fig. 1 consisting of three phase boundaries, the middle one being a bisector, is commonly seen in shape-memory alloys. It is one of only two possible arrangements which can occur if $\underset{\sim}{F}_1 = \underset{\sim}{1}$ and if $\underset{\sim}{F}_2$ and $\underset{\sim}{F}_3$ are crystallographically equivalent, in the sense that $\underset{\sim}{F}_2 = \underset{\sim}{Q}\underset{\sim}{F}_3\underset{\sim}{Q}^T$, $\underset{\approx}{Q}$ being a rotation.

An analysis of these arrangements is made easy by the following fact: an arrangement like that shown in Fig. 2 permits a coherent deformation if and only if any arrangement diffeomorphic to it permits a coherent deformation. A diffeomorphism also preserves the condition that each surface is a surface of non-zero discontinuity of the deformation gradient. Thus, to treat a variety of arrangements with curved surfaces, it is sufficient to analyze arrangements of planes meeting at points and at line segments.

According to equation (2.2), the arrangement of Fig. 2 only supports discontinuities (at $\underset{\sim}{x}$) with parallel amplitudes, a property shared by some of the simplest arrangements. For example, the two arrangements shown in Fig. 3, which both have six interfaces, also have parallel amplitudes at $\underset{\sim}{x}$.

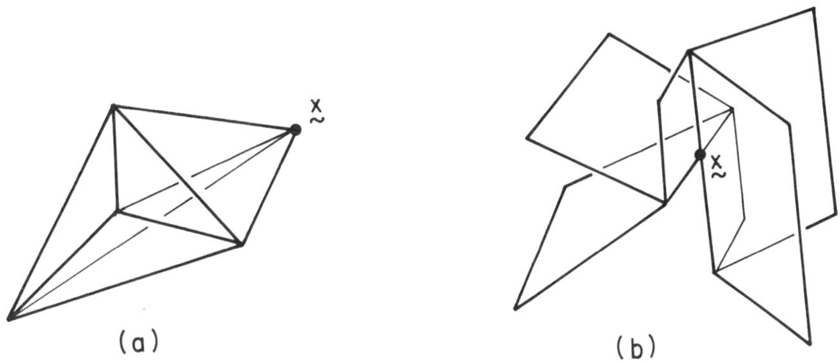

Fig. 3. Arrangements with parallel amplitudes at $\underset{\sim}{x}$.

These two arrangements (and arrangements diffeomorphic to them) represent the simplest arrangements of phases meeting at a point which do not consist of a number of leaves meeting on an axis (like the arrangement shown in

Fig. 2). Here, 'simplest' means that the smallest number of surfaces of nonzero discontinuity of the deformation gradient are present. See [1] for precise statements.

I now summarize some facts concerning the stability of coherent phases in a dead loading device. I shall assume the existence of a free energy function of the form

$$\phi(\underset{\sim}{F}, \theta), \tag{2.4}$$

$\underset{\sim}{F}$ being the deformation gradient and θ being the temperature. The Piola stress $\underset{\sim}{T}$ is defined by

$$\underset{\sim}{T} = \frac{\partial\phi}{\partial\underset{\sim}{F}}(\underset{\sim}{F},\theta); \tag{2.5}$$

it has the interpretation that

$$\underset{\sim}{t} = \underset{\sim}{T}\,\underset{\sim}{n} \tag{2.6}$$

is the force per unit referential area on a surface in the deformed configuration of the body whose inverse image in the reference configuration has a normal $\underset{\sim}{n}$.

If a body described by a free energy function $\phi(\underset{\sim}{F},\theta)$ is loaded by assigned dead loads $\underset{\sim}{t}(\underset{\sim}{x})$, $\underset{\sim}{x} \in \partial\mathcal{R}$ we are led to define the total energy functional by

$$\Phi[\mathcal{X}] = \int_{R} \phi(\nabla\underset{\sim}{\mathcal{X}}(\underset{\sim}{x}),\theta)\,dV - \int_{\partial\mathcal{R}} \underset{\sim}{t}(\underset{\sim}{x})\cdot\underset{\sim}{\mathcal{X}}(\underset{\sim}{x})\,dA \tag{2.7}$$

A <u>stable</u> deformation is one which minimizes Φ, by definition, with θ = const. Suppose a continuous, piecewise linear deformation given by

$$\begin{aligned} \underset{\sim}{y} = \tilde{\underset{\sim}{\mathcal{X}}}(\underset{\sim}{x}) &= \underset{\sim}{F}_i\underset{\sim}{x} + \underset{\sim}{c}_i, \qquad x \in \mathcal{R}_i, \\ \underset{\sim}{F}_i &= \text{const.}, \\ \underset{\sim}{c}_i &= \text{const.}, \; i = 1,\ldots,n, \\ \bigcup_i \mathcal{R}_i &= \mathcal{R} \end{aligned} \tag{2.8}$$

is stable. It follows from (2.1), (2.8) and the continuity of $\tilde{\underset{\sim}{\mathcal{X}}}$ that $\partial\mathcal{R}_i \cap \partial\mathcal{R}_j$ either has zero area or is a nonempty subset of a plane. If the latter is true, I shall say that $\mathcal{R}_i$ <u>borders</u> <u>on</u> $\mathcal{R}_j$.

Several properties of the minimizer $\tilde{\underset{\sim}{\mathcal{X}}}$ which follow directly from the fact that it is stable are well-known.

These are:

1) Equilibrium. If $\mathcal{R}_i$ borders on $\mathcal{R}_j$ and the dividing plane has a normal $\mathbf{n}_{ij}$, then

$$\left(\frac{\partial\phi}{\partial \mathbf{F}}(\tilde{\mathbf{F}}_i,\theta) - \frac{\partial\phi}{\partial \mathbf{F}}(\tilde{\mathbf{F}}_j,\theta)\right)\mathbf{n}_{ij} = \mathbf{0}. \tag{2.9}$$

2) Local stability of the interface. If $\mathcal{R}_i$ borders on $\mathcal{R}_j$, then*

$$\phi(\tilde{\mathbf{F}}_i,\theta) - \phi(\tilde{\mathbf{F}}_j,\theta) - (\tilde{\mathbf{F}}_i - \tilde{\mathbf{F}}_j)\cdot\frac{\partial\phi}{\partial \mathbf{F}}(\tilde{\mathbf{F}}_j,\theta) = 0. \tag{2.10}$$

3) Natural boundary conditions.

$$\frac{\partial\phi}{\partial \mathbf{F}}(\nabla\tilde{\mathbf{X}},\theta)\mathbf{n} = \mathbf{t} \quad \text{on } \partial\mathcal{R} \tag{2.11}$$

Derivations of (2.10) have been given by Abeyaratne [2], Gurtin [3] and James [4].

I am able to give a simple description of the piecewise linear minima $\tilde{\mathbf{X}}$ if I assume that the arrangement of the sets $\mathcal{R}_1,\ldots,\mathcal{R}_n$ is not too complicated. A slight generalization of Theorem 5 of reference [1] suggests the following:

Definition 1. A piecewise linear deformation $\tilde{\mathbf{X}}$ given by (2.8) is termed simple if for each $i \in \{1,\ldots,n\}$, each member of the set $\{\tilde{\mathbf{F}}_1,\ldots,\tilde{\mathbf{F}}_{i-1},\tilde{\mathbf{F}}_{i+1},\ldots,\tilde{\mathbf{F}}_n\}$ of deformation gradients is defined on some region which borders $\mathcal{R}_i$.

Clearly, a piecewise linear deformation is simple if each region borders on every other region, for example, as shown in Fig. 2 or Fig. 3a. The slight generalization of Theorem 5 of reference [1] referred to above is:

Theorem 1. Let a simple, continuous, piecewise linear deformation of the form (2.9) be stable. Then,

(i) the Piola stress is constant on all of $\mathcal{R}$:

$$\frac{\partial\phi}{\partial \mathbf{F}}(\tilde{\mathbf{F}}_i,\theta) = \mathbf{T} = \text{const.}; \tag{2.12}$$

---

* The dot product of two tensors is defined by $\mathbf{A}\cdot\mathbf{B} = \operatorname{tr}\mathbf{A}\mathbf{B}^T$.

(ii) $\underset{\sim}{F}_i$ is a point of convexity of $\phi(\cdot,\theta)$, i.e.

$$\phi(\underset{\sim}{F},\theta) - \phi(\underset{\sim}{F}_i,\theta) - (\underset{\sim}{F}-\underset{\sim}{F}_i) \cdot \underset{\sim}{T} \geqq 0 \quad \forall \underset{\sim}{F}. \quad (2.13)$$

The proof of this result uses the necessary condition (2.11) in an important way; it is possible to construct (not stable) piecewise linear deformations which satisfy equilibrium (2.9) but do not satisfy (2.12). Notice that, generally, the Cauchy or true stress is not the same in the various regions.

A glance at the proof of Theorem 5 of [1] yields a similar result; if it is assumed a priori that (2.12) holds for a continuous, piecewise linear deformation, then (2.13) is satisfied regardless of whether or not the deformation is simple. I think there are probably sufficiently complicated stable piecewise linear deformations which do not have a uniformly constant Piola stress.

As a matter of terminology, $\underset{\sim}{F}_i$ will be called a point of convexity of $\phi(\underset{\sim}{F},\theta)$ if it satisfies (2.13).

3. SOME OBSERVATIONS

A number of metals and minerals have behavior similar to that described in the Introduction. Here, I shall describe the behavior of certain 18R martensites which suggest some interesting theoretical problems. I found the work of Saburi, Wayman, Takata and Nenno [5] most useful, to which I refer the reader for more details. The behavior of the 18R martensites is similar to the behavior of 2H, 3R and 9R martensites, as discussed by Saburi and Wayman [6]. This includes about a dozen alloys. I shall discuss observations on Cu-Zn-Ga.

Single crystals of Cu-20.4% Zn-12.5% Ga(at .%) have cubic symmetry above a certain temperature θ_o. If the temperature is lowered to θ_o, needles and platelets of martensite grow into the crystal from its boundary. Each needle looks like a divided wedge and each contains two or four regions upon which apparently constant deformation gradients are defined. If the temperature is at θ_o, or just below θ_o, and the arrangement is not too complicated, the phase boundaries appear straight*.

In this case, it appears that the deformation is reasonably assumed to be continuous and piecewise linear. A typical observation might look (schematically) like Fig. 4.

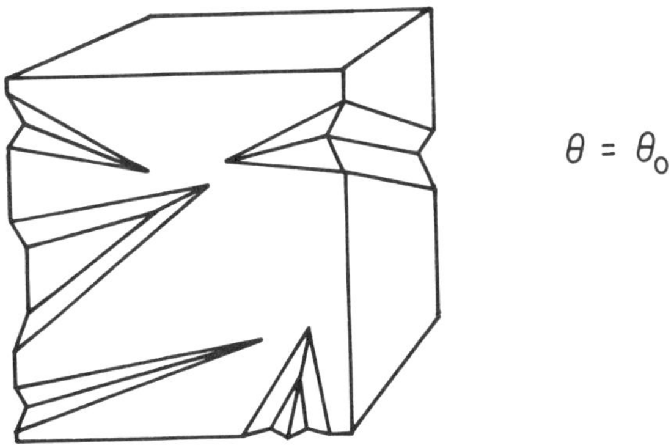

Fig. 4. Appearance of the specimen at the transformation temperature.

---

* Saburi and Wayman discuss in detail the fine structure of the martensite in reference [6]. The theory to be presented here is aimed at macroscopic behavior, at the scale of the optical microscope, so $\underset{\sim}{F}$ represents macroscopic change of shape. There seems to be no major obstacle to formulating a similar theory for the fine structure, in which $\underset{\sim}{F}$ represents the submicroscopic change of shape. If this were done for the alloy described above, the boundary between the parent phase and the internally twinned martensitic phase would actually be zig-zagged. This follows from the statements made just after our equation (2.3), and from the description of the change of shape in the fine structure offered by Saburi and Wayman.

A commonly observed arrangement of four variants is shown in Fig. 5.

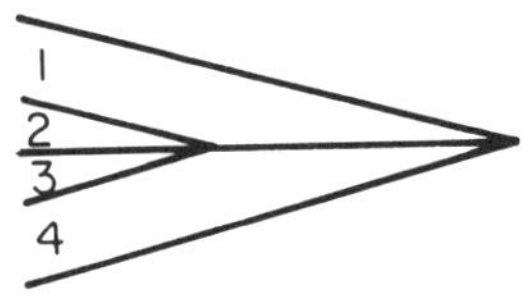

Fig. 5

It can be understood from the following calculation based on coherence. Expressed relative to a suitable orthonormal basis coincident with the crystallographic axes of the parent crystal, the deformation gradient in region 1 of Fig. 5 is $\underset{\sim}{F}_1 = \underset{\sim}{1} + \underset{\sim}{a}_1 \otimes \underset{\sim}{n}_1$, in which

$$\begin{aligned} \underset{\sim}{a}_1 &= \text{const.} \times (0,1,1), \\ \underset{\sim}{n}_1 &= (.14,.70,-.70). \end{aligned} \tag{3.1}$$

These are approximate relations; using the shape deformation matrices calculated by Saburi and Wayman [6, Table 5], I get $\underset{\sim}{a}_1 \alpha (.17,.84,1)$ and $\underset{\sim}{n}_1 = (.14,.70,-.70)$. These are calculated from a crystallographic theory of martensite based on kinematics; coherence calculations suggest that $\underset{\sim}{a}_1$ cannot differ this much from (0,1,1). The remaining deformation gradients are given by

$$\underset{\sim}{F}_i = \underset{\sim}{1} + \underset{\sim}{a}_i \otimes \underset{\sim}{n}_i = \underset{\sim}{R}_i \underset{\sim}{F}_1 \underset{\sim}{R}_i^T, \text{ no sum, } i=1,\ldots,4. \tag{3.2}$$

in which

$$\underset{\sim}{R}_2 = \begin{pmatrix} 1 & 0 & 0 \\ 0 & 0 & -1 \\ 0 & 1 & 0 \end{pmatrix}, \quad \underset{\sim}{R}_3 = \begin{pmatrix} 1 & 0 & 0 \\ 0 & -1 & 0 \\ 0 & 0 & -1 \end{pmatrix}, \quad \underset{\sim}{R}_4 = \begin{pmatrix} -1 & 0 & 0 \\ 0 & 0 & -1 \\ 0 & -1 & 0 \end{pmatrix} \tag{3.3}$$

A short calculation based on (3.2), (3.3) and (2.4) shows that the $\underset{\sim}{F}_i$ are gradients of a continuous deformation. Note that this coherence is possible only if $\underset{\sim}{a}_1$ lies on the axis of each of the rotations given in (3.3). The fact that $\underset{\sim}{a}_1$ lies on the axis of symmetry transformations of the parent phase leads to a rich variety of arrangements, since many arrangements have parallel amplitudes.

Generally, 24 distinct deformation gradients are observed, describing the deformation in each 'variant'. Let $\underset{\sim}{n}_i$ denote the unit normal of the plane which separates the parent phase and the $i^{\underline{th}}$ variant, counting $\pm\underset{\sim}{n}_i$ only once. The 24 normals obtained in this way are related by cubic symmetry transformations (rotations which map a cube into itself). Building on the notation introduced above, let deformation gradients be denoted by

$$\begin{aligned} \underset{\sim}{F}_i &= \underset{\sim}{1} + \underset{\sim}{a}_i \otimes \underset{\sim}{n}_i \\ &= \underset{\sim}{R}_i \underset{\sim}{F}_1 \underset{\sim}{R}_i^T \quad \text{(no sum)}, \quad i=1,\ldots,24. \end{aligned} \tag{3.4}$$

Under uniaxial extension, Saburi and Wayman observe that a single variant consumes the whole specimen, at least for the several initial orientations tested. The deformation gradient in the most favorably oriented variant in the loaded state is close (but not equal) to $\underset{\sim}{F}_i$, for some $i \in 1,\ldots,24$.

They show that the most favorably oriented variant is the one which produces the greatest extension in the tensile direction. This is consistent with a minimum energy criterion based upon (2.7) if the following conditions are assumed:

(a) All variants have the same free energy (This will follow from further restrictions imposed on ϕ in the next section).

(b) The elastic energy stored in each variant as it deforms under the load is sufficiently small.

(c) The loading is dead.

This follows because conditions b and c imply that the first term in (2.7) is the same for all variants, and the second term is minimized by the variant which yields the largest value of

$$\int_{\partial R} \underset{\sim}{t} \cdot \underset{\sim}{X} \, dA. \tag{3.5}$$

The theory presented here should be able to cope with the limitation imposed in b as well as more general loadings, stable configurations with more than one variant, and situations in which some of the unstressed parent or daughter phases have different free energies than others. The latter will naturally occur above the transformation temperature.

In the rest of this paper, I shall be concerned with conditions under which the parent phase can be obtained by loading at temperatures below the transformation temperature.

4. CONSTITUTIVE EQUATION

The macroscopic behavior of the alloy just described seems to be reasonably described by a free energy which depends upon the local change of shape and the temperature:

$$\phi(\underset{\sim}{F}, \theta). \tag{4.1}$$

The free energy is subject to the condition of Galilean invariance, i.e.

$$\phi(\underset{\sim}{Q}\underset{\sim}{F}, \theta) = \phi(\underset{\sim}{F}, \theta) \qquad \begin{array}{l} \underset{\sim}{Q} \in \sigma^+ \\ \underset{\sim}{F} \in \mathcal{D}, \end{array} \tag{4.2}$$

in which σ^+ is the group of all rotations and $\mathcal{D}$ is the domain of $\phi(\underset{\sim}{F}, \theta)$ with θ fixed. $\mathcal{D}$ is a subset of the tensors with positive determinants. I shall be concerned with a crystalline material, with which there is associated a finite group $\mathcal{P}^n$ of rotations of order n, known as the point group. The free energy shall satisfy

$$\phi(\underset{\sim}{F}\underset{\sim}{R}, \theta) = \phi(\underset{\sim}{F}, \theta) \qquad \begin{array}{l} \forall\, \underset{\sim}{R} \in \mathcal{P}^n \\ \forall\, \underset{\sim}{F} \in \mathcal{D}. \end{array} \tag{4.3}$$

For a material whose reference configuration is an undistorted cubic crystal, n = 24.

More general ideas of symmetry are clearly appropriate for some problems involving coherent phase transformations. These ideas of global symmetry are the subject of recent investigations by Ericksen [7] and Pitteri [8]. I seem to get by with the local notion of symmetry summarized by (4.3) if the phase boundaries are mobile and if $\underset{\sim}{F}$ represents the macroscopic change of shape. The deep reasons for this are not clear to me. I only note that in some cases the deformation of a phase as seen through an optical microscope appears adequately described by $\underset{\sim}{F}$, while the atomic deformation is a zig-zag on parallel planes a few atomic spacings apart, obviously only related to the macroscopic deformation gradient in an average sense. Also, it seems that one can restrict attention to a suitable subdomain of a globally defined free energy, if the energy barriers which restrict the body from undergoing transformations to points outside the subdomain are large. Further study would be enlightening.

A function satisfying (4.2) can always be expressed as a function of $\underset{\sim}{C} = \underset{\sim}{F}^T\underset{\sim}{F}$ only, i.e.

$$\phi(\underset{\sim}{F},\theta) = \tilde{\phi}(\underset{\sim}{C},\theta), \qquad \underset{\sim}{C} = \underset{\sim}{F}^T\underset{\sim}{F}. \tag{4.4}$$

In (4.1), $\underset{\sim}{F}$ is calculated from a certain reference configuration $\mathcal{R}$, which I shall assume is interpreted as the unstressed parent phase at the transformation temperature θ_o. If we keep the parent crystal unloaded and raise the temperature, it simply deforms homogeneously with a deformation gradient

$$\underset{\sim}{F}_P(\theta), \quad \theta \geqq \theta_o. \tag{4.5}$$

By our choice of reference configuration $\underset{\sim}{F}_P(\theta_o) = \underset{\sim}{1}$. Obviously $\underset{\sim}{F}_P$ is not uniquely determined, since a rigid rotation of the crystal at any value of $\theta \geqq \theta_o$ will not affect any of the given conditions. The parent phase appears stable at temperatures above θ_o, so it is reasonable to invoke a special case of Theorem 1. If we do so, we infer from (2.13) that

$$\phi(\underset{\sim}{F},\theta) \geqq \phi(\underset{\sim}{F}_P(\theta),\theta) \qquad \forall\, \underset{\sim}{F} \in \mathcal{D}, \quad \theta > \theta_o. \tag{4.6}$$

In terms of the function $\tilde{\phi}$, (4.6) becomes

$$\tilde{\phi}(\underset{\sim}{C},\theta) \geqq \tilde{\phi}(\underset{\sim}{C}_P(\theta),\theta), \tag{4.7}$$

in which $C_p = F_p^T F_p$. While it would be impossible to assume strict inequality in (4.6), in view of the condition (4.2), it is often reasonable to do so in (4.7) for $C \neq C_p$. One reason for doing this is that the strengthened (4.7), after being combined with (4.4) and (4.3), implies that

$$RC_p(\theta)R^T = C_p(\theta) \qquad \forall R \in P^n. \tag{4.8}$$

If P^{24} is the point group for a cubic crystal, it is known that the only positive-definite symmetric $C_p(\theta)$ satisfying (4.8) is a dilation, i.e.

$$C_p(\theta) = \alpha(\theta)1, \quad \alpha(\theta_o) = 1, \quad \theta \geqq \theta_o, \tag{4.9}$$

and (4.9) is consistent with observations on stable cubic crystals.

At $\theta = \theta_o$ the variants of the new phase appear, as discussed in §3. They do not generally correspond to a simple arrangement (compare Fig. 4 and Definition 1), although parts of the specimen do. In any case, if the arrangement is not too complicated, it appears from the flatness of phase boundaries that the parent phase is homogeneously deformed. Assuming this, and assuming that the parent phase meets ∂R at three points having linearly independent unit normal vectors, then we can show that the stress is everywhere zero in the parent phase. Assuming an arrangement like Fig. 4 and piecewise linear deformations, we can then prove that the stress is zero in the variants as well.

Having argued that the stress is zero everywhere, we now appeal to the remark made just after Theorem 1. From this we deduce that each F_i (associated with i^{th} variant) is a point of convexity of $\phi(F,\theta_o)$.

For $\theta > \theta_o$ the behavior of the specimen appears consistent with the idea that the unstressed parent phase loses stability. That is, if $\theta < \theta_o$ and θ is near θ_o there is no G near 1 having the property that the stress vanishes at G and that G is a point of convexity of $\phi(F,\theta)$. This does not forbid the possibility that the parent phase is stable in a stressed state for $\theta < \theta_o$.

I now summarize the constitutive assumptions described above. $\phi(\underset{\sim}{F},\theta)$ is a smooth function.

Summary

$\theta > \theta_o$ $\quad \underset{\sim}{C}_p(\theta)$ is the Cauchy-Green strain in the unstressed parent phase.

$$\tilde{\phi}(\underset{\sim}{C}_p(\theta),\theta) < \tilde{\phi}(\underset{\sim}{C},\theta) \quad \forall \underset{\sim}{C} \neq \underset{\sim}{C}_p(\theta), \tag{4.10}$$

$$\underset{\sim}{C}_p(\theta_o) = \underset{\sim}{1}. \tag{4.11}$$

$\theta = \theta_o$ $\quad \underset{\sim}{F}_i$ is the deformation gradient in the $i^{\underline{th}}$ unstressed variant, i=1,...,n. $\underset{\sim}{1}$ is the deformation gradient in the parent phase.

$$\underset{\sim}{F}_i = \underset{\sim}{1} + \underset{\sim}{R}_i\underset{\sim}{a} \otimes \underset{\sim}{R}_i\underset{\sim}{n}, \quad \underset{\sim}{R}_i \in P^n, \tag{4.12}$$

$$\tilde{\phi}(\underset{\sim}{F}_i^T\underset{\sim}{F}_i,\theta_o) = \tilde{\phi}(\underset{\sim}{1},\theta_o), \tag{4.13}$$

$$\tilde{\phi}(\underset{\sim}{1},\theta_o) < \tilde{\phi}(\underset{\sim}{C},\theta_o) \quad \forall \underset{\sim}{C} \neq 1, \underset{\sim}{F}_i^T\underset{\sim}{F}_i; \quad i = 1,\ldots,n. \tag{4.14}$$

$\theta < \theta_o$ $\quad \overline{\underset{\sim}{F}}_i(\theta)$ is the deformation gradient in the ith variant, $\overline{\underset{\sim}{F}}_i(\theta_o) = \underset{\sim}{F}_i$.

$$\overline{\underset{\sim}{F}}_i(\theta) = \underset{\sim}{R}_i\underset{\sim}{F}_1(\theta)\underset{\sim}{R}_i^T, \quad \underset{\sim}{R}_i \in P^n, \quad \underset{\sim}{R}_1 = \underset{\sim}{1}, \tag{4.15}$$

$$\tilde{\phi}(\overline{\underset{\sim}{F}}{}_i^T\overline{\underset{\sim}{F}}_i,\theta) < \tilde{\phi}(\underset{\sim}{C},\theta) \quad \forall \underset{\sim}{C} \neq \underset{\sim}{F}_k^T\underset{\sim}{F}_k, \quad k = 1,\ldots,n. \tag{4.16}$$

5. DESCRIPTION OF THE UNSTRESSED VARIANTS

The assumptions (4.13) and (4.14) say that each $\underset{\sim}{F}_i$ is a minimizer of $\phi(\underset{\sim}{F},\theta_o)$. Clearly, any deformation gradient of the form

$$\underset{\sim}{Q}\underset{\sim}{F}_i\underset{\sim}{R}_k, \quad \underset{\sim}{Q} \in \sigma^+, \; \underset{\sim}{R}_k \in P^n \tag{5.1}$$

also minimizes $\phi(\underset{\sim}{F},\theta_o)$, and therefore also yields zero stress. More generally, transformations of the form $\underset{\sim}{Q}\ldots\underset{\sim}{R}_k$ take points of convexity into points of convexity.

Then why are these other deformation gradients not seen as variants? The answer is that they are seen, in complicated arranements, but that they never border on the parent phase. To see this, let $\underset{\sim}{F}_k$ be one of deformation gradients given by (4.12). If a deformation gradient of the form (5.1) also borders coherently on the parent phase, we must have

$$\mathbf{Q}\mathbf{F}_i\mathbf{R}_k = \mathbf{1} + \mathbf{b} \otimes \mathbf{m}, \tag{5.2}$$

for some vectors $\mathbf{b}$ and $\mathbf{m}$. Written out, (5.2) becomes

$$\hat{\mathbf{Q}}(\mathbf{1} + \mathbf{a} \otimes \mathbf{n})\mathbf{R}_e = \mathbf{1} + \mathbf{b} \otimes \mathbf{m}, \tag{5.3}$$

in which $\mathbf{R}_e = \mathbf{R}_k\mathbf{R}_i$ and $\hat{\mathbf{Q}} = \mathbf{Q}\mathbf{R}_i$. $\mathbf{R}_e$ belongs to P^n because P^n is a group. Given $\mathbf{a}$ and $\mathbf{n}$, we wish to know if there are rotations $\mathbf{Q}$ and $\mathbf{R}_e \in P^n$ and vectors $\mathbf{b}$ and $\mathbf{m}$ which satisfy (5.3).

It is shown in [1] that if $\mathbf{a}$ is not parallel to $\mathbf{n}$, there are vectors $\hat{\mathbf{b}}$ and $\hat{\mathbf{m}}$ such that <u>every</u> solution $(\hat{\mathbf{b}},\hat{\mathbf{m}})$ of (5.3) can be written

$$\mathbf{b} = \pm \mathbf{R}_j\hat{\mathbf{b}}, \qquad \mathbf{m} = \pm \mathbf{R}_j\hat{\mathbf{m}}, \tag{5.4}$$

for some $\mathbf{R}_j \in P^n$. In (5.4) the ± signs are associated. Formulae for $\hat{\mathbf{b}}$ and $\hat{\mathbf{m}}$ are given in [1]. If $\mathbf{a}$ is parallel to $\mathbf{n}$, there are no solutions of (5.2).

In general, there is no simple relation between $(\hat{\mathbf{b}},\hat{\mathbf{m}})$ and $(\mathbf{a},\mathbf{n})$. However, for a number of alloys which undergo martensitic transformations, it happens that measured values of $\mathbf{a}$ and $\mathbf{n}$ yield crystallographically equivalent vectors $\hat{\mathbf{b}}$, $\hat{\mathbf{m}}$, <u>i.e.</u>

$$\hat{\mathbf{b}} = \mathbf{R}_p\mathbf{a}, \quad \hat{\mathbf{m}} = \mathbf{R}_p\mathbf{n} \quad \text{for some } \mathbf{R}_p \in P^n. \tag{5.5}$$

In this case, we simply recover the old set of variants from (5.3), pairs being related by a rigid rotation. This occurs for the alloy discussed in §3. Saburi and Wayman [6] use the notation n(+) and n(-), n=1,1',2,2',...,6,6' to denote pairs of variants related as in (5.5).

This can also be understood from the point of view of the crystallographic theory of martensitic transformations [6], where it is associated with the notion of "K-degeneracy". The alloy described in §3 has this property. If (5.5) is satisfied for a material with a given point group P^n, I shall say that the pair $(\mathbf{a},\mathbf{n})$ is <u>degenerate</u>.

If $(\mathbf{a},\mathbf{n})$ is degenerate, we expect to see at most n variants bordering on the parent phase, n being the order of P^n. Many martensitic transformations are degenerate. Also, many martensitic transformations have the property that $\underset{\sim}{n}$ is the normal to an "irrational plane", meaning that $\underset{\sim}{R}_k\underset{\sim}{n}$, k=1,...,n produces n distinct normals. This leads to the existence of exactly n variants. On the other hand, for non-degenerate $(\underset{\sim}{a},\underset{\sim}{n})$ we would expect to see up to 2n variants bordering on the parent phase, according to (5.4).

This all presumes that only a single phase is present, in the sense that all variants are described as invariant transformations of $\underset{\sim}{1} + \underset{\sim}{a} \otimes \underset{\sim}{n}$. One could make similar statements if two or more phases were present.

Notice that if a variant does not border on the parent phase, as in Fig. 5, it need not have a deformation gradient of the form $\underset{\sim}{1} + \underset{\sim}{R}_i\underset{\sim}{a} \otimes \underset{\sim}{R}_i\underset{\sim}{n}$.

6. IS THE PARENT PHASE STABLE UNDER DEAD LOADING?

It is of interest to decide whether the parent phase can be obtained by dead loading at temperatures less than or equal to θ_o. To simplify the notation in this Section, I shall put $\theta = \theta_o$. The convex hull of the points of convexity of $\phi(\underset{\sim}{F},\theta_o)$ plays an important role in calculations of this kind. To describe this set, let (orbit$\{\underset{\sim}{F}_i\}$) denote all tensors of the form $\underset{\sim}{Q}\underset{\sim}{F}_i\underset{\sim}{R}_k$ in which $\underset{\sim}{Q}$ is a rotation, $\underset{\sim}{F}_i$ is either given by (4.12) or equals $\underset{\sim}{1}$, and $\underset{\sim}{R}_k$ belongs to the point group. Let

$$\mathcal{H} = \text{convex hull (orbit}\{\underset{\sim}{F}_i\}). \tag{6.1}$$

The theory of convex sets (see Rockafellar [9], for example) shows that any $H \in \mathcal{H}$ can be written as a finite convex combination of members of orbit$\{\underset{\sim}{F}_i\}$, viz.,

$$\begin{aligned} &\underset{\sim}{H} \in \mathcal{H} \Rightarrow \underset{\sim}{H} = \sum_{k=1}^{m} \lambda_k \underset{\sim}{G}_k, \text{ where} \\ &\qquad \lambda_k \geqq 0, \\ &\qquad \underset{\sim}{G}_k \in \text{ orbit}\{\underset{\sim}{F}_i\}, \quad k=1,\dots,m \\ &\qquad \sum_{k=1}^{m} \lambda_k = 1. \end{aligned} \tag{6.2}$$

In fact, we can even choose m = 9 and still obtain all of $\mathcal{H}$, in the case under consideration.

The main interest in $\mathcal{H}$ stems from the following result.

Lemma 1. If $\underset{\sim}{H} \in \mathcal{H}$ and $\underset{\sim}{H}$ is a point of convexity of $\phi(\underset{\sim}{F},\theta_o)$, then the stress vanishes at $\underset{\sim}{H}$:

$$\frac{\partial\phi}{\partial \underset{\sim}{F}}(\underset{\sim}{H},\theta_o) = \underset{\sim}{0} \tag{6.3}$$

Proof. Let $\underset{\sim}{H}$ be given by (6.2) and let $\underset{\sim}{H}$ be a point of convexity of $\phi(\underset{\sim}{F},\theta_o)$. Then, in particular

$$\phi(\underset{\sim}{G}_k,\theta_o) - \phi(\underset{\sim}{H},\theta_o) - (\underset{\sim}{G}_k-\underset{\sim}{H}) \cdot \frac{\partial\phi}{\partial \underset{\sim}{F}}(\underset{\sim}{H},\theta_o) \geqq 0. \tag{6.4}$$

Multiply (6.4) by λ_k and sum from k=1 to k=m, using (6.2):

$$\Sigma\lambda_k\phi(\underset{\sim}{G}_k,\theta_o) - \phi(\underset{\sim}{H},\theta_o) - (\Sigma\lambda_k\underset{\sim}{G}_k-\underset{\sim}{H}) \cdot \frac{\partial\phi}{\partial \underset{\sim}{F}}(\underset{\sim}{H},\theta_o) \geqq 0. \tag{6.5}$$

If we now use (6.2) again together with the fact that $\phi(\underset{\sim}{G}_k,\theta_o)$ is independent of k, we obtain from (6.5),

$$\phi(\underset{\sim}{G}_k,\theta_o) \geqq \phi(\underset{\sim}{H},\theta_o). \tag{6.6}$$

But since $\underset{\sim}{G}_k \in \text{orbit}\{\underset{\sim}{F}_i\}$ and since the $\underset{\sim}{F}_i$ minimize $\phi(\underset{\sim}{F},\theta_o)$, the $\underset{\sim}{G}_k$ must also minimize $\phi(\underset{\sim}{F},\theta_o)$; by (6.6), $\underset{\sim}{H}$ also minimizes $\phi(\underset{\sim}{F},\theta_o)$. Therefore, the derivative of $\phi(\underset{\sim}{F},\theta_o)$ with respect to $\underset{\sim}{F}$, evaluated at $\underset{\sim}{H}$, vanishes.□

Thus, all deformation gradients belonging to simple, stable, continuous, loaded, piecewise linear deformations will lie outside of $\mathcal{H}$.

By the definition of $\mathcal{H}$, $\underset{\sim}{1} \in \mathcal{H}$. Suppose that $\underset{\sim}{1}$ does not lie on $\partial\mathcal{H}$. Then, there are no tensors sufficiently close to $\underset{\sim}{1}$ which lie outside of $\mathcal{H}$. In this sense the parent phase (i.e. a deformation gradient near $\underset{\sim}{1}$) could not be recovered by applying dead loads at $\theta = \theta_o$.

My first impression was that $\mathbf{1}$ would not belong to ∂H if the order n of P^n were sufficiently large. This impression was based on (4.12), which suggests that the tensors $\mathbf{F}_i$ in some sense "surround" $\mathbf{1}$. The fact is: even if n is arbitrarily large, there are cases in which $\mathbf{1} \in \partial H$.

To see this, consider the expression (6.2) for an arbitrary $\mathbf{H} \in H$. For any tensor $\mathbf{F}$, let $||\mathbf{F}|| = (\mathrm{tr}\, \mathbf{F}^T\mathbf{F})^{\frac{1}{2}}$. Since $\lambda_k \geqq 0$,

$$||\mathbf{H}|| = ||\Sigma\lambda_k\mathbf{G}_k|| \leqq \Sigma\lambda_k||\mathbf{G}_k||, \tag{6.7}$$

and

$$\begin{aligned} ||\mathbf{G}_k||^2 &= \mathrm{tr}(\mathbf{R}_i^T\mathbf{F}_k^T\mathbf{Q}^T\mathbf{Q}\mathbf{F}_k\mathbf{R}_i) \\ &= \mathrm{tr}((\mathbf{1} + \mathbf{n} \otimes \mathbf{a})(\mathbf{1} + \mathbf{a} \otimes \mathbf{n})) \\ &= 3 + 2\mathbf{a}\cdot\mathbf{n} + |\mathbf{a}|^2, \end{aligned} \tag{6.8}$$

which is independent of k. Thus, by combining $(6.8)_3$, (6.7) and $\Sigma\lambda_k = 1$, we get

$$||\mathbf{H}|| \leqq (3 + 2\mathbf{a}\cdot\mathbf{n} + |\mathbf{a}|^2)^{\frac{1}{2}}. \tag{6.9}$$

However, $||\mathbf{1}|| = \sqrt{3}$. Suppose $2\mathbf{a}\cdot\mathbf{n} + |\mathbf{a}|^2 \leqq 0$. Then, by (6.9) each member of H has a norm less than or equal to $\sqrt{3}$. Clearly, there are tensors arbitrarily close to $\mathbf{1}$ with a norm greater than $\sqrt{3}$. Thus, we have,

<u>Lemma 2</u>. <u>If</u> $2\mathbf{a}\cdot\mathbf{n} + |\mathbf{a}|^2 \leqq 0$, <u>then</u>*

$$\mathbf{1} \in \partial H. \tag{6.11}$$

This Lemma does not definitely establish that there are $\mathbf{G}$ near $\mathbf{1}$ which are points of convexity of $\phi(\mathbf{F},\theta_o)$, but this seems likely. I have carried out the analysis of a simplified problem in which there are no invariance groups. In this simplified case, the points of convexity of the free energy lie in cones which emanate from ∂H. The invariance groups complicate the matter considerably, but a complete analysis should be feasible. Some rough calculations suggest that the arrangement of phases will be determined by a delicate relation among the vectors $\mathbf{a}$ and $\mathbf{n}$, the structure of P^n, and the elastic moduli of the various phases.

---

* Lemma 2 does not apply to the alloy discussed in §3 because $\mathbf{a}\cdot\mathbf{n} \approx 0$ for it. Probably it is the case that $\mathbf{1} \notin \partial H$ for this alloy.

REFERENCES

1. R. D. James, The mechanics of coherent phase transformations in solids. Brown University Technical Report, October, 1982.
2. R. Abeyaratne, An admissibility criterion for equilibrium shocks in finite elasticity. J. Elasticity 13 (1983) P. 175-184.
3. M. E. Gurtin, Two phase deformations of elastic solids, to appear.
4. R. D. James, Finite deformation by mechanical twinning. Arch. Rational Mech. Anal. 77 (1981), p. 143-176.
5. T. Saburi, C. M. Wayman, K. Takata and S. Nenno, The shape memory mechanism in 18R martensitic alloys. Acta. Met. 28 (1980), p. 15-32.
6. T. Saburi and C. M. Wayman, Crystallographic similarities in shape memory martensites. Acta. Met. 27 (1979), p. 979-995.
7. J. L. Ericksen, The Cauchy and Born hypotheses for crystals, Tech. Summary Report, Mathematics Research Center, University of Wisconsin, Madison, Wisconsin. October, 1983.
8. M. Pitteri, Reconciliation of local and global symmetries of crystals, to apper in J. Elasticity.
9. R. Tyrell Rockafellar, Convex Analysis. Princeton University Press (1970).

This work was partially supported by the National Science Foundation under the grant MEA-8209303 and the Materials Research Laboratory at Brown University.

Division of Engineering
Brown University
Providence, RI 02912

THE GRADIENT THEORY OF PHASE TRANSITIONS ON A FINITE INTERVAL

M. E. Gurtin

1. Introduction.

In a paper [1][1] now classic, van der Waals considered fluids whose free energy at constant temperature is determined not only by the density, but also by the density gradient. Cahn and Hilliard [3], apparently unaware of van der Waals paper, rederived what is essentially van der Waals' theory and, using this theory, obtained several important results concerning interfacial energy between phases. Since then, gradient theories have been used to analyze phase transitions, spinodal decomposition, and other physical phenomena (cf. [2, 4] for selected references).

In van der Waals' theory the energy of a vessel of unit cross section, extending from $x = -L$ to $x = L$, is

$$E_\varepsilon(\rho) = \int_{-L}^{L} [W(\rho(x)) + \varepsilon^2 \rho'(x)^2] dx \quad . \tag{1.1}$$

Here $\rho(x)$ is the fluid density, $W(\rho)$ is the (free) energy per unit volume, and $\varepsilon > 0$ is a small parameter. If the total mass in the container is M, then we have the additional constraint

$$\int_{-L}^{L} \rho(x) dx = M \quad . \tag{1.2}$$

[1] Cf. the translation by Rowlinson [2].

ISBN 0-12-309770-3

Van der Waals, following Gibbs, believed that the stable configurations of the fluid are those which minimize (1.1) subject to (1.2). In modern terminology, this suggests the problem:

(P_ε) minimize (1.1) over all $\rho \in H^1(-L,L)$, $\rho > 0$,

which satisfy the constraint (1.2) .

Here $H^1(-L,L)$ is the usual Sobolev space of square-integrable functions possessing square-integrable generalized derivatives.

For W sufficiently regular, the direct method of the calculus of variations and elementary regularity theory lead to the conclusion[1] that Problem P_ε possesses a (not necessarily unique) solution, so existence is not at issue here. The goal instead is to identify the minimizers of P_ε when the "chemical potential" $W'(\rho)$ has the form illustrated in Figure 1, a form motivated by the original potential of van der Waals. In this paper I shall discuss recent work of Jack Carr, Marshall Slemrod, and myself [6] concerning this problem.[2]

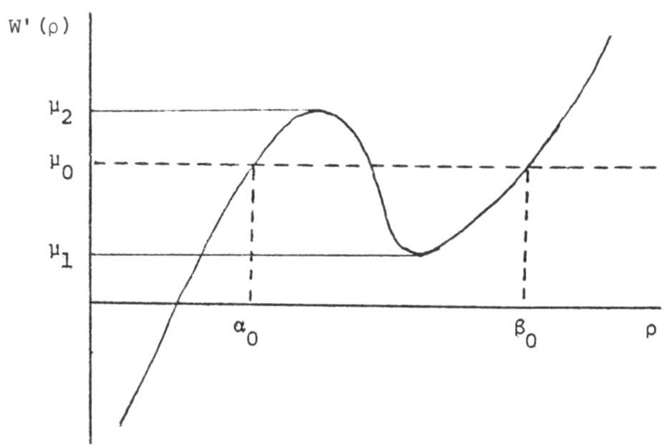

Figure 1. Chemical potential $W'(\rho)$ as a function of density ρ.

---

[1] Cf. Morrey [5], Theorems 1.9.1 and 1.10.1.

[2] There is a large and rapidly growing literature - which we make no attempt to discuss - on problems with $L = \infty$. In this connection cf., e.g., Aifantes and Serrin [7], Cahn and Hilliard [3], Coleman [8], and Davis and Scriven [4].

2. <u>The problem without structure</u> ($\varepsilon = 0$).

Consider first the problem with $\varepsilon = 0$, for which (1.1) has the form

$$E_0(\rho) = \int_{-L}^{L} W(\rho(x))dx . \tag{2.1}$$

This problem may be stated as follows:

(P_0) <u>minimize</u> (2.1) - <u>subject to</u> (1.2) - <u>over all</u> $\rho > 0$ <u>with</u> $\rho, W(\rho) \in L^1(-L,L)$.

P_0 is easily solved with the aid of the auxiliary functional

$$\int_{-L}^{L} [W(\rho(x)) - \mu\rho(x)]dx$$

in which μ(= constant) is a Lagrange multiplier. For a minimum to exist the Euler-Lagrange equation and Weierstrass-Erdmann corner conditions must be satisfied; i.e.,

$$\begin{aligned} &W'(\rho) = \mu \text{ at points of continuity of } \rho, \text{ while} \\ &W(\rho) - \mu\rho \quad \text{is continuous across jumps in } \rho . \end{aligned} \tag{2.2}$$

Inspection of (2.2) shows that solutions are either constant (<u>single phase</u>) or piecewise constant (<u>two phase</u>); and in the latter case have the form

$$\rho_0(x) = \begin{cases} \alpha_0, & x \in S_1 \\ \beta_0, & x \in S_2 \end{cases} \tag{2.3}$$

with S_1, S_2 disjoint measurable sets whose union is $[-L,L]$, and with a_0, β_0, and μ_0 defined by the <u>Maxwell conditions</u> (cf. Figure 1)

$$W(\beta_0) - W(\alpha_0) = \mu_0(\beta_0-\alpha_0)$$

$$\mu_0 = W'(\alpha_0) = W'(\beta_0) .$$

Further, letting

$$\ell_i = \text{measure}(S_i) ,$$

(1.2) yields

$$\ell_1 = \frac{2L(\beta_0-r)}{\beta_0 - \alpha_0} , \quad \ell_2 = \frac{2L(r-\alpha_0)}{\beta_0 - \alpha_0} , \quad r = \frac{M}{2L} ; \tag{2.4}$$

and since $\ell_i > 0$, a necessary condition for the existence of a two-phase solution is that the <u>average density</u> r

satisfy

$$\alpha_0 < r < \beta_0 . \tag{2.5}$$

When (2.5) is satisfied, any $\rho_0(x)$ of the form (2.3) with ℓ_i given by (2.4) is a global minimizer for Problem P_0; and the corresponding energy is

$$E_0(\rho_0) = 2L[W(\alpha_0) + \mu_0(r-\alpha_0)] . \tag{2.6}$$

If $r \leqslant \alpha_0$ or $r \geqslant \beta_0$ the above discussion shows that a two-phase solution of P_0 is impossible; here the minimizer is simply the single-phase solution

$$\rho(x) \equiv r .$$

3. The problem with structure $\varepsilon > 0$.

As noted in Section 2, for $\alpha_0 < r < \beta_0$ there are two-phase solutions of P_0; in fact, there is an uncountable infinity of such solutions. Here we shall attempt to answer the question:

Are any of the two-phase solutions (2.3), (2.4) - in some physical sense - preferred?

Indeed, the theory with $\varepsilon = 0$ allows the formation of interfaces (jumps in density) without a concomitant increase in energy. One might expect that in a theory which includes interfacial energy, the two-phase solutions with least energy would be the single-interface solutions

$$\rho_0(x) = \begin{Bmatrix} \alpha_0, & -L \leqslant x \leqslant -L + \ell_1 \\ \beta_0, & -L + \ell_1 < x \leqslant L \end{Bmatrix} \quad \text{and} \quad \rho_0(-x) . \tag{3.1}$$

As shown in [6], this expectation is, in fact, justified. To explain the results of [6], consider the theory with $\varepsilon > 0$. While this theory does not allow for jumps in density, it does allow ρ to suffer rapid changes over small intervals, and such changes are penalized in energy by the term $\varepsilon^2(\rho')^2$ in (1.1). Thus the theory with $\varepsilon > 0$ has associated with it a natural interfacial energy.

Theorem 1. Let $r \in (\alpha_0, \beta_0)$. Then:

(i) for small $\varepsilon > 0$ and modulo reversals,[1] Problem P_ε has a unique global minimizer $\rho_\varepsilon(x)$;

(ii) $\rho_\varepsilon(x)$ is strictly monotone;

(iii) as $\varepsilon \to 0$, $\rho_\varepsilon(x)$ (or its reversal) approaches the single-interface solution $\rho_0(x)$

Thus the single-interface solutions (3.1) are preferred in the following sense: they represent limits, as $\varepsilon \to 0$, of solutions within the van der Waals theory $(\varepsilon > 0)$, a theory which penalizes rapid changes in density.

We define the interfacial energy[2] σ_ε to be the difference

$$\sigma_\varepsilon = E_\varepsilon(\rho_\varepsilon) - E_0(\rho_0)$$

between the actual energy and the energy (2.6) of the global minimizer when $\varepsilon = 0$. Cahn and Hilliard [3] have shown that the interfacial energy for a medium of infinite extent is $\varepsilon\sigma_0$ with σ_0 the constant

$$\sigma_0 = 2 \int_{\alpha_0}^{\beta_0} [W(\xi) - W(\beta_0) - \mu(\xi-\beta_0)]^{1/2} d\xi .$$

Theorem 2. Let $r \in (\alpha_0, \beta_0)$. Then the interfacial energy σ_ε has the asymptotic form

$$\sigma_\varepsilon = \varepsilon\sigma_0 + O(e^{-C/\varepsilon})$$

as $\varepsilon \to 0$, with $C > 0$.

Remark. The range $r \notin (\alpha_0, \beta_0)$ is uninteresting: there (as for P_0) all solutions of P_ε are constant.

The proof of Theorem 1 is based on a systematic study of the associated system consisting of the Euler-Lagrange equation, the natural boundary conditions, and the constraint:[3]

---

[1] $f(-x)$ is the reversal of $f(x)$.

[2] Cf. [3].

[3] Cf. Novick-Cohen and Segel [9], who obtain a solution of S_ε in terms of Jacoby elliptic functions for $W(\rho)$ quartic in ρ.

$$S_\varepsilon \begin{cases} 2\varepsilon^2\rho'' = W'(\rho) - \mu , \\ \rho'(\pm L) = 0 , \\ \int_{-L}^{L} \rho(x)dx = M . \end{cases}$$

One step in the proof consists in showing that monotonic solutions of S_ε have lower energy than nonmonotonic solutions. In fact:

Theorem 3. Nonmonotonic solutions of S_ε are unstable in the sense that they cannot be even local minimizers for P_ε.

In the next two sections we will give the main ideas underlying the proofs of Theorems 1 and 2, and we will prove Theorem 3. All of these results are taken from [6], where complete proofs can be found.

4. Alternative formulation in terms of integral equations.

If we let

$$x = \varepsilon t, \; z(t) = \rho(\varepsilon t) \tag{4.1}$$

and, for convenience, write

$$a = \frac{L}{\varepsilon} ,$$

then S_ε takes the form

$$2\ddot{z} = W'(z) - \mu , \tag{4.2}$$

$$\dot{z}(\pm a) = 0 , \tag{4.3}$$

$$\int_{-a}^{a} z(t)dt = 2ar , \tag{4.4}$$

with r the average density (2.4). Equation (4.2) has the first integral

$$\dot{z}^2 = \Phi_\mu(z) - b \tag{4.5}$$

with b constant and

$$\Phi_\mu(z) = W(z) - \mu z$$

the associated Gibbs function (Figure 2). Note that, by (4.3), the boundary values

$$z_1 = z(-a), \; z_2 = z(a) \tag{4.6}$$

of $z(t)$ are solutions y of the equation

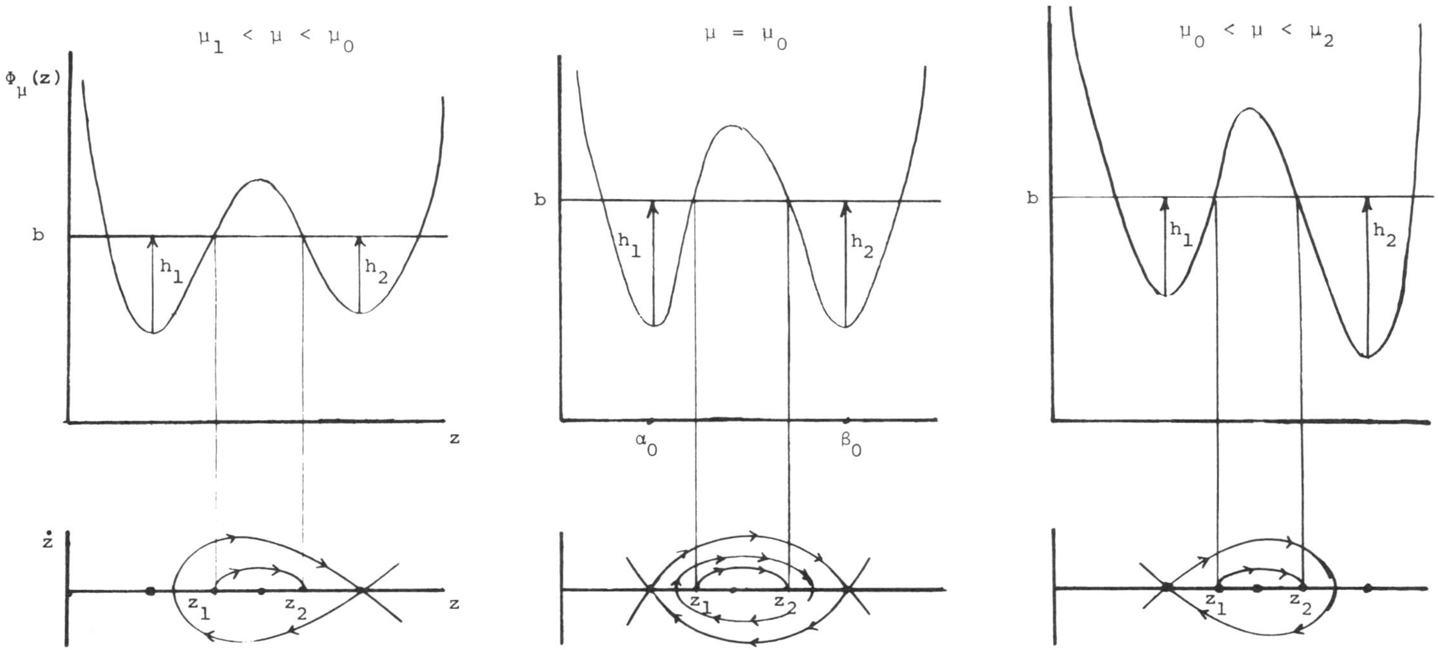

Figure 2. The Gibbs function and the phase portraits, the points $z_i = z_i(\Delta)$, and the parameters h_i.

$\Phi_\mu(y) = b$.

Finding a nonconstant solution of (4.2) and (4.3) is equivalent to finding a trajectory-segment (cf. the phase portraits in Figure 2) that begins and ends on the axis $\dot{z} = 0$ and has duration $2a$. Such segments must necessarily lie on periodic orbits, and these, in turn, are possible only when $\mu \in (\mu_1, \mu_2)$ (cf. Figure 1); we therefore restrict our attention to this interval.

Let

$$\Delta = (\mu, b) .$$

It is clear from Figure 2 that for each $\mu \in (\mu_1, \mu_2)$ there is a range of values of b for which $\Phi_\mu(y) = b$ at four distinct values of y; we label the inner two values by $z_1(\Delta)$ and $z_2(\Delta)$, $z_1(\Delta) < z_2(\Delta)$, we refer to such Δ as admissible.

Let us agree to use the term simple solution for a nonconstant solution $z(t)$ of (4.2) - (4.4) with $z(t) \geq 0$. For $z(t)$ simple we may use (4.5), (4.6), and the change in variable $\xi = z(t)$ to convert the integral

$$\int_{-a}^{a} z(t)^n dt$$

to one of the form

$$I_n(\Delta) = \int_{z_1(\Delta)}^{z_2(\Delta)} \xi^n [\Phi_\mu(\xi) - b]^{-1/2} d\xi . \tag{4.7}$$

Thus, since $n = 0$ yields the duration of $z(t)$, while $n = 1$ leads to the constraint integral in (4.4), we have the integral equations

$$I_0(\Delta) = 2a, \quad I_1(\Delta) = 2ar . \tag{4.8}$$

Using as a basis the above argument, it is not difficult to show that there is a one-to-one correspondence between simple solutions and admissible Δ that satisfy (4.8):

Theorem 3. Given any simple solution, the corresponding Δ satisfies (4.8); conversely, given any admissible solution Δ of (4.8), there exists exactly one simple solution to which Δ corresponds.

The solution $\rho_\varepsilon(x)$ of S_ε - mentioned in Theorem 1 - is obtained by solving (4.8) near μ_0. The analysis is quite complicated: the main ideas are as follows.

The chief contribution to the integral $I_0(\Delta)$ comes from the singularities at $\xi = z_1(\Delta), z_2(\Delta)$. If $T_1(\Delta)$ and $T_2(\Delta)$ denote the "contributions" from $z_1(\Delta)$ and $z_2(\Delta)$ to $I_0(\Delta)$, then

$$I_0(\Delta) \sim T_1(\Delta) + T_2(\Delta) .$$

Since the integrand in $I_1(\Delta)$ weights these singularities by $z_1(\Delta)$ and $z_1(\Delta)$, while $z_1(\Delta) \sim \alpha_0$ and $z_2(\Delta) \sim \beta_0$,

$$I_1(\Delta) \sim \alpha_0 T_1(\Delta) + \beta_0 T_2(\Delta) .$$

Thus to satisfy (4.8),

$$T_1(\Delta) \sim a\ell_1, \quad T_2(\Delta) \sim a\ell_2 \tag{4.9}$$

with ℓ_i given by (2.4). Clearly, for consistency in (4.9) we must have (2.5), a condition that plays a crucial role in the analysis for $\varepsilon = 0$.

To solve (4.9) and hence (4.8), we need to know how $T_i(\Delta)$ depends on Δ. Here we give only the approximate results:

$$T_i(\Delta) \sim -B_i \ell n h_i , \tag{4.10}$$

where

$$h_1 = b - \varphi_1(\mu), \quad h_2 = b - \varphi_2(\mu) ,$$

with $\varphi_i(\mu)$ the values assumed by the Gibbs function Φ_μ at its two minima (cf. Figure 2), while

$$B_1 = [2W''(\alpha_0)]^{-1/2}, \quad B_2 = [2W''(\beta_0)]^{-1/2} .$$

If we substitute (4.10) into (4.9) and solve for h_i, we find that

$$h_i \sim u_i \exp\left(-\frac{C_i}{\varepsilon}\right)$$

with u_i and $C_i > 0$ constants independent of $\varepsilon (= L/a)$.

This crude calculation motivates introducing new parameters $k = (k_1, k_2)$ defined implicitly by the scaling

$$h_i = \exp\left(k_i - \frac{C_i}{\varepsilon}\right) .$$

When this is done (4.8) takes the form

$$k = F(k, \varepsilon) , \tag{4.11}$$

where $F = (F_1, F_2)$ is a C^1 function whose derivative with respect to k vanishes at $\varepsilon = 0$. Hence we can solve (4.11) for $k = k(\varepsilon)$ by the implicit function theorem.

The above argument proves that for ε small enough and $\alpha_0 < r < \beta_0$, (4.2) - 4.4) has a solution $z(t)$ which is close to the heteroclinic orbit.

After descaling we obtain a solution $\rho_\varepsilon(x)$ of S_ε. Using the above properties of $z(t)$, we are able to show that $\rho_\varepsilon(x)$ obeys (ii) and (iii) of Theorem 1. In the next section we will sketch our argument showing that $\rho_\varepsilon(x)$ is the global minimizer for P_ε.

5. Energy.

In the analysis sketched in Section 4 we established the existence of a strictly increasing solution $\rho_\varepsilon(x)$ of S_ε with μ close to μ_0. To show that $\rho_\varepsilon(x)$ is a global minimizer, we must show that:

(1) $\rho_\varepsilon(x)$ has lower energy than constant solutions;

(2) $\rho_\varepsilon(x)$ has lower energy than strictly-increasing solutions of S_ε with μ bounded away from μ_0;

(3) $\rho_\varepsilon(x)$ has lower energy than nonmonotonic solutions of S_ε.

(We have tacitly used the fact that the energy is invariant under reversals.)

In view of the variable change (4.1), the energy (1.1) has the form

$$E_\varepsilon(z) = \varepsilon \int_{-a}^{a} [W(z(t)) + z(t)^2]dt \ . \tag{5.1}$$

Consider first (1) and (2). The fact that strictly-increasing solutions of S_ε correspond to simple solutions of (4.2) - (4.4) allows us to utilize the formulation in terms of admissible Δ. Indeed, for $z(t)$ simple we can use (4.5) to rewrite the functional $E_\varepsilon(z)$ as a function $E_\varepsilon(\Delta)$ of the admissible $\Delta = (\mu, b)$ corresponding to $z(t)$; using (4.4), the result is

$$E_\varepsilon(\Delta) = 2L(\mu r+b) + 2\varepsilon \int_{z_1(\Delta)}^{z_2(\Delta)} [\Phi_\mu(\xi)-b]^{1/2}\, d\xi \ .$$

Remark. The critcal points of $E_\varepsilon(\Delta)$ are exactly the admissible solutions of the integral equations (4.8).

Our proofs of (1), (2), and Theorem 2 are based on a systematic study of the function $E_\varepsilon(\Delta)$. The final assertion, (3), is an obvious consequence of Theorem 3, which we now prove.

Proof of Theorem 3. Let z be a nonmonotonic solution. We will show that z cannot be a local minimizer by showing that the second variation

$$\frac{d^2}{d\xi^2} E_\varepsilon(z+\xi\eta)\Big|_{\xi=0} \tag{5.2}$$

is strictly negative at some $\eta \in H^1(-a,a)$ with

$$\int_{-a}^{a} \eta(t)dt = 0 . \tag{5.3}$$

Write $\varepsilon J(z,\eta)$ for the second variation (5.2); then (5.1) yields

$$J(z,\eta) = \int_{-a}^{a} [2\dot{\eta}^2 + W''(z)\eta^2]dt . \tag{5.4}$$

Since nonconstant solutions of $S_{\varepsilon r}$ lie on closed orbits, and since z is nonmonotonic, z must repeat: there is a $T \in (-a,a]$ such that

$$z(T) = z(-a), \quad \dot{z}(T) = 0 \ (= \dot{z}(-a)) .$$

Let

$$\eta_0(t) = \begin{cases} \dot{z}(t), & -a \leq t \leq T \\ 0 , & T \leq t \leq a , \end{cases}$$

let $\eta_1(t)$ be any $H^1(-a,a)$ function that satisfies (5.3) and

$$\eta_1(-a) = 1, \quad \eta_1(t) = 0 \quad \text{for} \quad T \leq t \leq a ,$$

and let

$$\eta(t) = \eta_0(t) + \kappa\eta_1(t)$$

with κ to be determined. Then $\eta \in H^1(-a,a)$ and satisfies (5.3). If we substitute η into (5.4) and use the fact that, by (4.2), $2\dddot{z} = W''(z)\dot{z}$, we find that

$$J(z,\eta) = -4\kappa\ddot{z}(-a) + O(\kappa^2) .$$

Since z is not constant, $\ddot{z}(-a) \neq 0$; thus we can choose κ such that $J(z,\eta) < 0$.

6. Application to solids.

The theory discussed thus far has possible application to the study of phase transformations in elastic solids. Consider a one-dimensional elastic filament which occupies the interval $-L \leq x \leq L$ in a fixed reference.

Using as a basis van der Waals' argument, Carr, Gurtin, and Slemrod [10] considered a stored-energy function of the form

$$W(\rho) + \varepsilon(\rho')^2$$

with ρ the strain and W' of the form shown in Figure 1. (A general theory of elastic materials with stored energy dependent on strain gradients was proposed by Toupin [11]; Ericksen [12] has suggested a nonconvex W as a model for "martensitic" materials.) Within this framework, the constraint (1.2) is the requirement that the filament have length M in the deformed configuration, and Problem $\mathcal{P}_\varepsilon$ corresponds to the hard-loading device of Ericksen [12] in which the displacement is specified at the ends of the filament.[1] The soft device, in which the force is specified, is analyzed in detail in [10].

References

[1] van der Waals, J. D., The thermodynamic theory of capillarity under the hypothesis of a continuous variation of density (in Dutch), Verhandel, Konink. Akad. Weten. Amsterdam (Sect. 1) Vol. 1, No. 8 (1893).

[2] Rowlinson, J. S., Translation of J. D. van der Waals' "The thermodynamic theory of capillarity under the hypothesis of a continuous variation of density", J. Stat. Phys. **20**, 197-244 (1979).

---

1 Cf. Antman [13, 14] and Antman and Carbone [15], who establish - for a class of hyperelastic rods in tension - the existence of solutions of the type that describe necking. Within our theory a neck would correspond to a complete loop in phase space and as such would be unstable (Theorem 3).

[3] Cahn, J. W. and J. E. Hilliard, Free energy of a non-uniform system. I. Interfacial free energy, J. Chem. Phys. 28, 258-267 (1958).

[4] Davis, H. T. and L. E. Scriven, Stress and structure in fluid interfaces, Advances Chem. Phys. 49, 357-454 (1982).

[5] Morrey, C. B., Multiple Integrals in the Calculus of Variations, Springer-Verlag, Berlin (1966).

[6] Carr, J., M. E. Gurtin, and M. Slemrod, Structured phase transformations on a finite interval, Arch. Rational Mech. Anal. To appear.

[7] Aifantes, E. D. and J. B. Serrin, Towards a mechanical theory of phase transformations, Technical Report, Corrosion Research Center, University of Minnesota, Minneapolis, MN (1980).

[8] Coleman, B. D., Necking and drawing in polymeric fibers under tension, Arch. Rational Mech. Anal. 83, 115-137 (1983).

[9] Novick-Cohen, A. and L. A. Segal, Nonlinear aspects of the Cahn-Hilliard equation. Forthcoming.

[10] Carr, J., M. E. Gurtin, and M. Slemrod, One-dimensional structured phase transformations under prescribed loads, Technical Report 2559, Mathematics Research Center, University of Wisconsin, Madison, WI (1983). J. Elast. to appear.

[11] Toupin, R. A., Elastic materials with couple-stresses, Arch. Rational Mech. Anal. 11, 385-414 (1962).

[12] Ericksen, J. L., Equilibrium of bars, J. Elasticity 5, 191-201 (1975).

[13] Antman, S. S., Nonuniqueness of equilibrium states for bars in tension, J. Math. Anal. Appl. **44**, 333-349 (1973).

[14] Antman, S. S., Qualitative theory of the ordinary differential equations of nonlinear elasticity, Mechanics Today, 1972.

[15] Antman, S. S. and E. R. Carbone, Shear and necking instabilities in nonlinear elasticity, J. Elast. **7**, 125-151 (1977).

Department of Mathematics
Carnegie-Mellon University
Pittsburgh, PA 15213

ONE-DIMENSIONAL SHOCK LAYERS IN KORTEWEG FLUIDS

R. Hagan and J. Serrin

In an earlier paper [3] the authors have studied the question of shock layers connecting equilibrium states in a van der Waals fluid. In certain respects this work parallels the classical theory of Gilbarg [2] for the case of fluids with convex adiabats (the perfect gas being the archetypical example of this situation), but there are two crucial differences. First, a van der Waals fluid exhibits regions of state space for which the fluid can be in a liquid phase, and second, the appropriate dynamic form of the stress tensor involves terms (first introduced by Korteweg) in which density gradients induce extra stresses. Because of the first difference it is possible to have a shock layer in which the fluid, before the arrival of the shock, is in its vapor phase, but, after passage of the shock, is compressed to a liquid phase. Such shock layers can be considered as dynamic phase transitions.

Our purpose in the present paper is to extend the results of [3] to quite general equations of state, including both van der Waals fluids and perfect gases as particular special cases. We are thus able to provide a comprehensive theory which takes into accoutn not only the well-known case of classical shock layers, but also allows for the appearance of general vapor-liquid phase transitions.

ISBN 0-12-309770-3

The basic equations which we shall use are those of continuum fluid mechanics, modified by an appropriate interstitial working term to model the energetic effect of the Korteweg stress. This modification, as proposed first in [1] , has the particularly useful property that it supplies a Liapunov function for the shock layer problem, providing thereby a mechanism for obtaining the required orbit connection between the two equilibrium end states.

Our main results include proof that the end states of a shock layer connection must satisfy the classical Rankine-Hugoniot conditions. What is equally interesting, however, is the converse fact that not all states satisfying these conditions can be connected by shock layers.

Other papers related to the present one are [4] and [5]. The first of these studies dynamic phase transitions near isothermal conditions, a situation quite different from the highly temperature dependent supersonic transitions which we find. The second deals only with Navier-Stokes type stress, and moreover restricts to equations of state having no spinodal set (or to domains of the state space outside the spinodal region), but within these limitations obtains somewhat similar existence theorems.

The experimental results described in [8] concerning supersonic phase transitions apparently fall into the shock layer regime studied here, at least when complete liquefaction occurs behind the shock.

1. Basic equations for shock layer connections

We consider steady one-dimensional flow of a general Korteweg fluid. Letting the speed of the flow at the location x, $-\infty < x < \infty$, be denoted by $u(x)$, the density by $\rho(x)$, and the temperature by $\theta(x)$, we see from the equation of continuity that ρu = constant. We suppose throughout that this constant is non-zero, in order to model dynamic transitions. Moreover, without loss of generality it may be taken positive. We thus have the basic relation

$$\rho u = m$$

valid for $-\infty < x < \infty$, where m is a fixed positive constant.

We seek flows in which $u(x)$, $\rho(x)$, $\theta(x)$ tend to well-defined limits (u_0,ρ_0,θ_0) and (u_1,ρ_1,θ_1) as x tends respectively to $-\infty$ and $+\infty$. Naturally we interpret (u_0,ρ_0,θ_0) to be the upstream fluid state, before the fluid has encountered any change of state, and (u_1,ρ_1,θ_1) as the downstream fluid state, after passing through some transition region. Such a flow, assuming it to exist at all, is called a shock layer when $\rho_1 \neq \rho_0$. As we shall see, the limit triples (u_0,ρ_0,θ_0) and (u_1,ρ_1,θ_1) must satisfy the Rankine-Hugoniot conditions if any shock layer connection at all is to exist.

The appropriate differential equations of motion for shock layer connections were derived in [3]. They are based on the following set of constitutive relations, see [1, Section 1]:

$$\begin{aligned} T = &(-p + \lambda \operatorname{div} \vec{v})I + 2\mu D \\ &+ (\alpha\Delta\rho + \beta|\operatorname{grad}\rho|^2 + \gamma \operatorname{grad}\rho \cdot \operatorname{grad}\theta)I \\ &+ \delta \operatorname{grad}\rho \otimes \operatorname{grad}\rho , \end{aligned}$$

$$q = \kappa \operatorname{grad}\theta$$

$$k = c\dot{\rho} \operatorname{grad}\rho .$$

Here T is the stress tensor, q the heat flux vector, and k the interstitial working. The first line in the formula for stress is the standard Navier-Stokes relation, in which p is the pressure, λ and μ the viscosity co-efficients, $\vec{v}$ the fluid velocity and D the rate of deformation tensor. The second line is the added stress due to density gradients, in line with Korteweg's theory. The heat flux is given by the standard Fourier law, with heat conduction coefficient κ. The coefficient c in the interstitial working is called the surface tension.

We assume throughout that the constitutive variables $(p,\lambda,\mu;\ \alpha,\beta,\gamma,\delta;\ \kappa;\ c)$ are given C^1 functions of ρ and θ, which are subject to the thermodynamic restrictions (see [1])

$$\mu > 0, \qquad 3\lambda + 2\mu \geqslant 0, \qquad \kappa > 0$$

$$\alpha = \rho c, \quad \beta = \frac{1}{2}\frac{\partial}{\partial\rho}(\rho c), \quad \gamma = \frac{\partial}{\partial\theta}(\rho c), \quad \delta = -c\ .$$

Further thermodynamic variables, ε = internal energy and η = entropy, are given by the formulae

$$\varepsilon = \tilde{\varepsilon}(\rho,\theta) + \frac{1}{2\rho}\left(c - \theta\frac{\partial c}{\partial\theta}\right)|\text{grad }\rho|^2$$

$$\eta = \tilde{\eta}(\rho,\theta) - \frac{1}{2\rho}\frac{\partial c}{\partial\theta}|\text{grad }\rho|^2\ .$$

Here $\tilde{\varepsilon}$ and $\tilde{\eta}$ are the equilibrium internal energy and equilibrium entropy related to the pressure by the Gibbs identity

$$\theta\, d\tilde{\eta} = d\tilde{\varepsilon} + p\, d\upsilon, \qquad \upsilon = \frac{1}{\rho}\ .$$

It is worthwhile observing that the corresponding free energy $\psi = \varepsilon - \theta\eta$ is just

$$\tilde{\psi}(\rho,\theta) + \frac{1}{2}\frac{c}{\rho}|\text{grad }\rho|^2\ ,$$

a formula originally proposed by van der Waals in his classical study of phase transition theory.

With these relations in hand, the equations governing the shock layer transition can now be stated. For convenience we introduce the specific volume $\upsilon = 1/\rho$ as the basic variable instead of ρ, and use a prime to denote derivatives with respect to x. Then we obtain the fundamental shock layer equations

$$\upsilon' = w$$

$$\sigma w' = m(\lambda + 2\mu)w - \frac{1}{2}\frac{\partial\sigma}{\partial\upsilon}w^2 - \frac{\partial\sigma}{\partial\theta}w\theta' - L(\upsilon,\theta)$$

$$\kappa\theta' = -\frac{m}{2}\frac{\partial}{\partial\theta}(\theta\sigma)w^2 + mM(\upsilon,\theta)\ .$$

The relation $w \equiv \upsilon'$ is introduced in order to obtain a first order system, rather than having to deal with a second order equation for υ coupled with a first order equation for θ. The thermodynamic functions L and M are given by

$$L(\upsilon,\theta) = p(\upsilon,\theta) + m^2(\upsilon - a)$$

$$M(\upsilon,\theta) = \tilde{\varepsilon}(\upsilon,\theta) - \frac{1}{2}m^2(\upsilon - a)^2\ ,$$

where the constant a is chosen so that $L(\upsilon_0,\theta_0) = 0$ and the additive constant in the equilibrium energy $\tilde{\varepsilon}$ is picked so that $M(\upsilon_0, \theta_0) = 0$. Finally we have put

$$\sigma = \rho^3 c .$$

It is worth noting that when the above equations are specialized to the case $c = 0$ they become simply

$$m(\lambda + 2\mu)\upsilon' = L(\upsilon, \theta)$$

$$\frac{\kappa}{m}\theta' = M(\upsilon, \theta) ,$$

the basic equations of classical shock layer theory (see [2]).

Since we are dealing with an autonomous system of equations it is clear that the prescribed end conditions can be met only if the states (u_0, ρ_0, θ_0) and (u_1, ρ_1, θ_1) are singular points of the system. It follows in turn that

$$\upsilon'(x),\ w'(x),\ \theta'(x) \to 0 \quad \text{as} \quad x \to \pm\infty .$$

Finally, we have the relation $u = m\upsilon$, which shows that the specific volume can equally be considered as a scaled representative of the speed u.

In addition to the differential equations of motion, one must also observe the Clausius-Duhem inequality. In the present circumstances this takes the form

$$m\,\eta' \geqslant \left(\frac{q}{\theta}\right)' ,$$

or alternately

$$\left(m\eta - \kappa\,\frac{\theta'}{\theta}\right)' \geqslant 0 .$$

Since $\theta' = 0$ at $x = \pm\infty$, it follows in particular that the front state entropy η_0 is less than or equal to the back state entropy η_1. Moreover $\eta_0 = \tilde{\eta}_0$ and $\eta_1 = \tilde{\eta}_1$ since $\upsilon' = 0$ at $\pm\infty$. Thus we have

$$\tilde{\eta}_1 \geqslant \tilde{\eta}_0 .$$

In fact only the inequality sign can hold. To see this, put

$$\Phi = \eta - \frac{\kappa}{m}\frac{\theta'}{\theta} = \frac{1}{\theta}\left(\theta\tilde{\eta} - M + \frac{1}{2}\sigma w^2\right) .$$

Then by a straightforward but tedious calculation we get

$$\Phi' = \frac{m}{\theta}(\lambda + 2\mu)\upsilon'^2 + \frac{\kappa}{m}\frac{\theta'^2}{\theta^2} .$$

Since $\Phi' \geqslant 0$ and since $\upsilon' > 0$ at least for some values x, the assertion is proved.

2. The Rankine-Hugoniot conditions.

For a shock layer connection to exist it is necessary to have $L(\upsilon_1,\theta_1) = M(\upsilon_1,\theta_1) = 0$ along with the originally noted conditions $L(\upsilon_0,\theta_0) = M(\upsilon_0,\theta_0) = 0$. This gives the relations

$$\rho_1 u_1 = \rho_0 u_0 = m$$

$$p_1 + \rho_1 u_1^2 = p_0 + \rho_0 u_0^2 = am^2$$

$$\tilde{\varepsilon}_1 - \frac{1}{2}(u_1 - am)^2 = \tilde{\varepsilon}_0 - \frac{1}{2}(u_0 - am)^2$$

which can be recognized immediately as the Rankine-Hugoniot end state conditions for a normal shock in a fluid with pressure p and energy $\tilde{\varepsilon}$.

If we write

$$H = M + \frac{1}{2}(\upsilon - \upsilon_0)L$$

then obviously $H = 0$ at both end states. An easy calculation shows that

$$H = \tilde{\varepsilon} - \tilde{\varepsilon}_0 + \frac{1}{2}(p + \rho_0)(\upsilon - \upsilon_0) \ ,$$

so H can be calculated solely from knowledge of the thermodynamic functions $\partial\tilde{\varepsilon}/\partial\theta$ and p together with the thermodynamic state (ρ_0,θ_0) in front of the shock.

In particular, if we make use of the variables (p,υ) then the back state (p_1,υ_1) can be found by intersecting the Hugoniot curve

$$H = 0$$

in the (p,υ) plane by the straight line

$$L = 0 \ ,$$

the latter having slope $-m^2$ and passing through the front state (ρ_0,υ_0). See Figure 1.

Figure 1

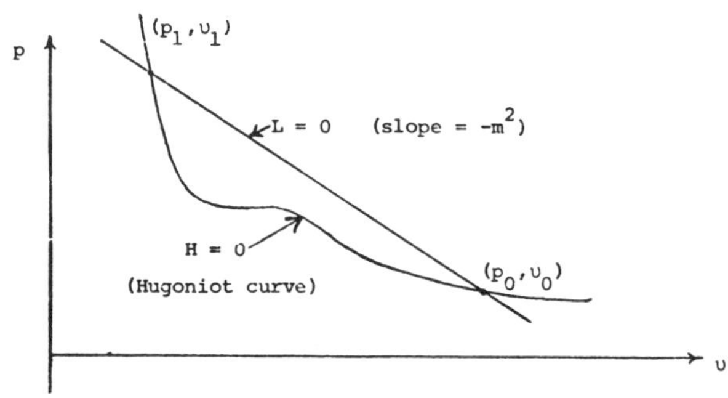

For this construction to succeed, that is, for there to exist a back state compatible with a given front state (p_0, υ_0) and mass flux m, it is of course necessary that certain conditions be met. To begin with, the change of variables $(\upsilon, \theta) \to (p, \upsilon)$ must be globally well-defined. To this end we make the hypothesis

$$\frac{\partial p}{\partial \theta} > 0 \,, \qquad \text{(H1)}$$

a natural one in fluid mechanics; and we ask furthermore that the state space (for simplicity) be of the form

$$\theta > 0, \qquad \upsilon > b$$

for some constant $b \geqslant 0$.

We shall assume in addition that along adiabats ($\tilde{\eta}$ = constant) in the (p, υ) plane the pressure tends to $+\infty$ as one proceeds to the left. Here it should be observed that

$$\left.\frac{\partial \tilde{\eta}}{\partial p}\right|_{\upsilon} = \frac{\partial \tilde{\eta}}{\partial \theta} \bigg/ \frac{\partial p}{\partial \theta} = \frac{1}{\theta}\, \frac{\partial \tilde{\varepsilon}}{\partial \theta} \bigg/ \frac{\partial p}{\partial \theta}$$

and that the right side quantity is positive under the following natural assumption

$$\frac{\partial \tilde{\varepsilon}}{\partial \theta} > 0 \,. \qquad \text{(H2)}$$

In consequence, these adiabats have single valued projections on the υ axis and may be continued throughout the state space until they reach one of the boundaries $\theta = 0$ or $\theta = \infty$ or $\upsilon = b$. Thus, by our assumptions, the family of adiabats in the (p, υ) plane forms a foliation of the state region, each adiabat having a single valued projection on the υ axis and each having an asymptote at some value $\upsilon = \hat{b} \geqslant b$. (The value $\hat{b}$ need not be the same for each adiabat, however.)

Under these minimal conditions it can be shown that any ray through (p, υ) which is directed into the second quadrant, and which initially lies above the curve $H = 0$, must eventually intersect $H = 0$ at a second point (p_1, υ_1). The latter point of course satisfies the conditions

$$p_1 > p_0, \qquad \upsilon_1 < \upsilon_0 \,.$$

These facts provide sufficient information to construct appropriate end state pairs (u_0, ρ_0, θ_0) and (u_1, ρ_1, θ_1)

allowing shock layer connections for Korteweg fluids. In particular, for fluids with convex adiabats the construction reduces to the standard one (cf. [2] and [6]), and provides all possible connections. For non-convex adiabats the situation is not quite so simple. Nevertheless interesting and useful results can still be obtained, as we shall show in the next section.

3. Existence of shock layer connections.

We shall say that a state $Z_0 = (p_0, \upsilon_0)$ in the (p,υ) plane is front side admissible for a shock layer connection provided that

(i) $\frac{\partial}{\partial \upsilon} p(\theta,\upsilon) < 0$ at Z_0,

and

(ii) the entropy $\tilde{\eta}$, considered as a function of (p,υ), is monotonically decreasing along each ray ℓ emanting from Z_0 and directed into the (open) sector indicated in Figure 2.

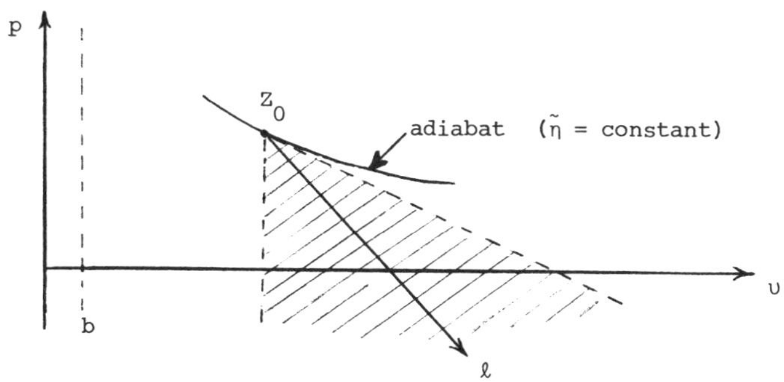

Figure 2

Condition (i) is the standard thermodynamical condition that the equilibrium state Z_0 should be stable. Condition (ii) has no simple thermomechanical interpretation, but is certainly satisfied for extensive regions of the state space of typical fluids, particularly at densities corresponding to gas phases and supercritical phases.

In addition to hypotheses (H1) and (H2) and the thermodynamic restrictions listed in Section 1, in what follows we shall also assume that

$$c > 0 , \qquad \text{(H3)}$$

$$\frac{\tilde{\varepsilon}}{\upsilon^2} \to 0 \quad \text{as} \quad \upsilon \to \infty . \qquad \text{(H4)}$$

Condition (H3) is a natural one in Korteweg's theory, corresponding to positive surface tension. Similarly condition (H4) holds for essentially all equations of state of practical interest.*

Theorem 1. Assume that the thermodynamic hypotheses listed above hold for a given Korteweg fluid. Let Z_0 be a front side admissible state and let u_0 be an assigned speed which is supersonic with respect to the state Z_0, that is

$$u_0 > a_0 ,$$

where $a^2 = \partial p/\partial \rho|_{\tilde{\eta}}$.

Then there exists a compression shock layer for the fluid, with front state Z_0 and front speed u_0, having moreover the following properties:

(i) the back state (u_1, ρ_1, θ_1) satisfies the Rankine-Hugoniot conditions relative to the front state (u_0, ρ_0, θ_0),

(ii) the back speed u_1 is subsonic with respect to the state $Z_1 = (p_1, \upsilon_1)$; in addition

$$p_1 > p_0, \quad \rho_1 > \rho_0, \quad \theta_1 > \theta_0, \quad \tilde{\eta}_1 > \tilde{\eta}_0 ,$$

$$\upsilon_1 < \upsilon_0, \quad u_1 < u_0 ,$$

(iii) no shock layer connection is possible for a back state satisfying any one of the following conditions

---

* Actually we need somewhat less, namely that

$$\overline{\lim_{\upsilon \to \infty}} \frac{\tilde{\varepsilon}(\upsilon, \theta)}{\upsilon^2} \le 0 \qquad \text{(uniformly on any interval } (0, \bar{\theta})),$$

a conclusion which holds for example whenever $\tilde{\varepsilon}$ is bounded above as when θ is bounded.

$$p < p_1, \quad \rho < \rho_1,$$
$$\upsilon > \upsilon_1, \quad u > u_1 \quad .$$

The proof of this result is similar to the proof of the corresponding Theorem 4.1 in [3], the principal tools used being the LaSalle Invariance Theorem and the fact that the function Φ introduced in Section 1 serves as a Liapunov function for the governing equations. Some of the arguments used in [3] moreover need to be altered since we are here dealing with a general equation of state rather than the special case of van der Waals equation (a full proof of Theorem 1 will be published elsewhere.)

The thermodynamic state Z_1 behind the shock can be constructed by drawing, on the Hugoniot diagram, a straight line of slope $-m^2$ through Z_0, where $m = \rho_0 u_0$. Because u_0 is supersonic with respect to the state Z_0 this line crosses the Hugoniot curve at Z_0 in such a way that immediately to the left of Z_0 it is above this curve and immediately to the right of Z_0 it is below. By the sector condition (ii) for a front side admissible state the line will not intersect the Hugoniot curve anywhere to the right of Z_0. On the other hand, as has already been noted in Section 2, this line will necessarily intersect $H = 0$ at some point to the left of Z_0. If there is only one such intersection, say at (p_1, υ_1), we take this for the state Z_1, and choose $u_1 = m\upsilon_1$. If there is more than one such intersection, we choose Z_1 to be the <u>nearest</u> of these to Z_0, and again take $u_1 = m\upsilon_1$.

This situation is illustrated in Figures 3 and 4, the shaded portions of the Hugoniot curve corresponding to states Z_1 which are attained through the above procedure. Those back states Z_1 which can occur in the procedure outlined above can be delineated explicitly by a "sun-ray" construction, indicated in Figure 5. To make the construction more evident, we have illustrated it for a Hugoniot curve with several loops, although in practice we know of no cases where more than one loop would appear.

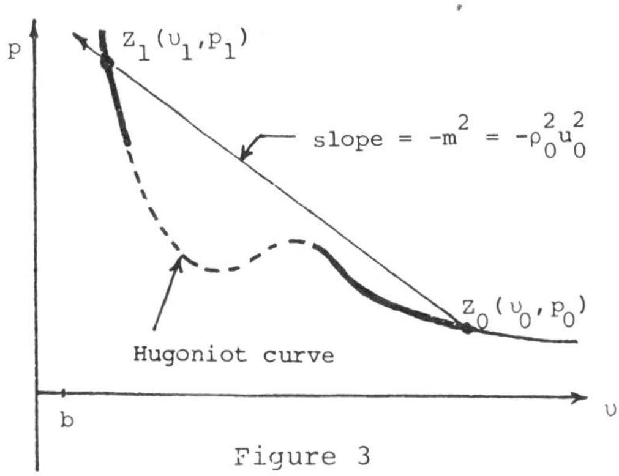

Figure 3

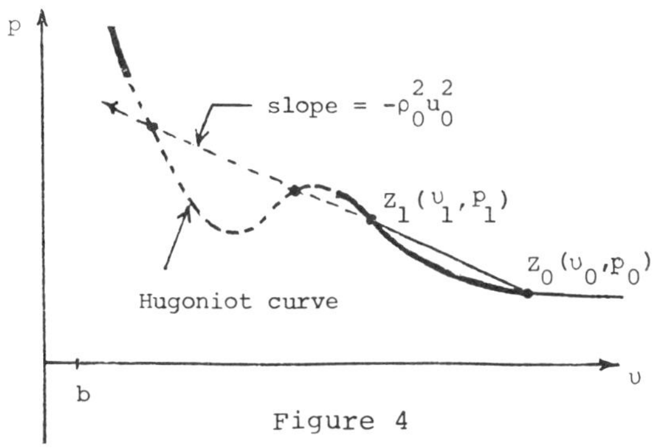

Figure 4

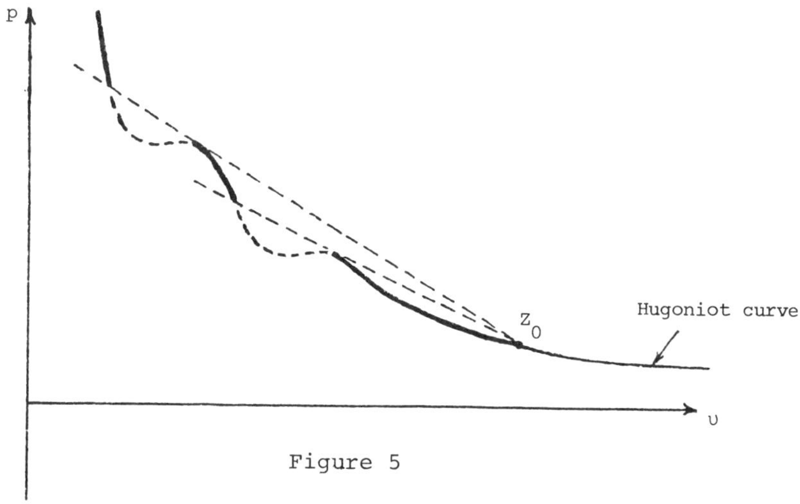

Figure 5

In addition to the main theorem above, the following result shows that back states to the right of Z_0 on the Hugoniot curve are also generally not accessible.

Theorem 2. Let the hypotheses of Theorem 1 hold. Then no shock layer connection can exist between the front state Z_0 and the front speed u_0, and any back state Z_1 to the right of Z_0.

The proof of Theorem 2 is essentially contained in the discussion immediately following the statement of Theorem 1. Theorem 2 can be stated alternately in the form that, under the hypotheses of Theorem 1, no rarefaction shock layer is possible for the front state Z_0 and front speed u_0.

When the adiabats in the (p,υ) plane are convex Gilbarg has shown that the part of the Hugoniot curve to the left of Z_0 is starlike with respect to Z_0. Thus the situation shown in Figure 4 does not occur in that case. Moreover, when the adiabats are convex the entropy $\tilde{\eta}$ on the Hugoniot curve to the right of Z_0 does not exceed $\tilde{\eta}_0$, so that shock connections are also not possible when Z_1 lies on this portion of the Hugoniot curve. Thus when adiabats are convex Theorem 1 supplies all possible shock layer connections.

For van der Waals fluids, and a fortiori for general equations of state, the Hugoniot curve can have loops such as those shown in Figure 3 and 4 (see [3] for a full discussion of this situation). In this case there are states Z_1 on the Hugoniot curve for which the present theorems give no information, e.g., in Figure 4, the two points of intersection of the straight line with the Hugoniot curve which lie beyond the indicated point Z_1.

Whether such points can actually be reached by a shock layer connection from Z_0 is an open question in general. We can however supply a partial answer in case c is small, or has the limiting value $c = 0$. Indeed in the latter case $(c=0)$ the governing equations reduce to those already discussed by Gilbarg, and a relatively straightforward extension of his methods shows that only the state Z_1

nearest to Z_0 can ever be reached. That is, when $c = 0$ all states on the dotted portions of the Hugoniot curves in Figure 3 and 4 are inaccessible to shock layer connections from Z_0.* A perturbation analysis, based on this result, can then be used to show that for sufficiently small positive values of the surface tension coefficient the same result is true, namely the dotted portions of the Hugoniot curve are inaccessible. Naturally such an analysis does not supply useful estimates of the particular size of c in question, and of course, even more, the particular values will themselves depend on how close a state Z_1 on the dotted portion of the Hugoniot curve is to neighboring (shaded) accessible states.

In line with the discussion in the preceding paragraph, one may ask whether the connections given by Theorem 1 are affected by having the other coefficients κ and $\lambda + 2\mu$ vanish. In outline form, the answers are as follows:

1. $\sigma = 0$. Connections continue to exist. They are monotone in the density and in the temperature, and can be considered as extensions to general fluids of the classical shock layer theory of Gilbarg [2].

2. $\kappa = 0$. A connection continues to exist for the same back states as in Theorem 1. No connection is possible to the right of Z_0.

3. $\mu = \lambda = 0$. A connection continues to exist for the same back states as in Theorem 1. No connection is possible to the right of Z_0.

4. $\kappa = \mu = \lambda = 0$, $c > 0$. No connections are possible between a front side admissible state Z_0 and any of the states Z_1 previously allowed (or to the right of Z_0). That states to the right of Z_0 remain inaccessible is obtained exactly as before. To show that states which were previously accessible (shaded parts of the Hugoniot curve)

---

* This holds even without the hypotheses (H3) and (H4) and even if the assigned speed u_0 is not supersonic.

are now inaccessible we consider the function Φ introduced in Section 1. Then because $\kappa = \lambda + 2\mu = 0$ we get

$$\Phi' = 0 .$$

Accordingly, any possible connection would have $\tilde{\eta}_1 = \tilde{\eta}_0$. On the other hand, an independent analysis of the Hugoniot curve shows that $\tilde{\eta}_1 > \tilde{\eta}_0$ at all the states Z_1 on the shaded part of the curve. Therefore such connections cannot exist.

Remark. The limit cases, $\sigma > 0$, $\lambda + 2\mu = 0$ on the one hand, and $\sigma = 0$, $\lambda + 2\mu > 0$ on the other, provide an interesting contrast when we compare equilibrium phase transitions and dynamic phase transitions. In particular, when $\sigma > 0$ and $\lambda + 2\mu = 0$ no shock layer $(m>0)$ can exist, as we have just noted, while conversely equilibrium transitions are always possible (see [7]). On the other hand, when $\sigma = 0$ and $\lambda + 2\mu > 0$, no equilibrium transition layers $(m=0)$ can exist, while Gilbarg's theory always yields dynamic transitions.

References

[1] E. Dunn and J. Serrin, On the thermodynamics of interstitial working. Institute for Mathematics and its Applications, Minneapolis. IMA Preprint 24. 1983.

[2] D. Gilbarg, The existence and limit behavior of the one-dimensional shock layer. American Journal of Mathematics, 7 (1951), 256-274.

[3] R. Hagan and J. Serrin. Dynamic changes of phase in a van der Waals fluid. Mathematics Research Center, University of Wisconsin-Madison. Technical Summary Report, 1984. To appear, New Perspectives in Thermodynamics, Springer-Verlag, 1984.

[4] R. Pego, Nonexistence of a shock layer in gas dynamics with a non-convex equation of state. Mathematics Research Center, University of Wisconsin-Madison. Technical Summary Report, 1983.

[5] M. Slemrod, Dynamic phase transitions in a van der Waals fluid. J. Differential Equations, 51 (1984).

[6] J. Serrin, Mathematical Principles of Classical Fluid Dynamics. Handbuch der Physik, vol. 8/1, Springer-Verlag, 1957. Especially Sections 56 and 57.

[7] J. Serrin, Phase transitions and interfacial layers for van der Waals fluids. Recent methods in Nonlinear Analysis and Applications, Camforo, Rionero, Sbordone, Trombetti, editors. Naples, Liguori Editori, 1981.

[8] P. A. Thompson, G. Dettleff, G. E. A. Meier and H.-D. Speckman, An experimental study of liquefaction shock waves. Journal of Fluid Mechanics, 95 (1979), 279-304.

Department of Mathematics
University of Oregon
Eugene, OR 97403

School of Mathematics
University of Minnesota
Minneapolis, MN 55455

DYNAMICS OF DENDRITIC PATTERN FORMATION

J. S. Langer

The material presented in this lecture is based largely on a brief review of modern developments in the theory of dendritic solidification prepared for publication in a metallurgical journal. The following text is essentially identical to that which appears there.[1]

The problem of dendritic pattern formation has roots in the field of metallurgy where some understanding of the unstable behavior of solidification fronts is essential for the interpretation and control of microstructures in alloys. A key advance was the work of Mullins and Sekerka[2,3] from which it became clear that morphological instabilities are intrinsically kinetic rather than thermodynamic in nature. Since then, progress has been slow.[4] The Mullins-Sekerka analysis can be viewed as doing for solidification theory approximately what Rayleigh and Chandrasekhar[5] did in identifying the onset of Benard convection patterns; both compute the way in which the quiescent state of a system becomes linearly unstable against infinitesimally weak pattern-forming deformations. Neither probes the far more interesting and challenging question of how structures emerge in these destabilized systems. For whatever reason, hydrodynamicists and applied mathematicians have pushed their version of the problem further than the metallurgists. With mathematical rigor and experimental precision, they have looked at nonlinear behavior near the onset of instabilities and have discovered a wealth of important phenomena. Much of the analogous work has yet to be done in solidification.

[1]From *Materials Science and Engineering*, Volume 65, pp. 37-44 (1984),

ISBN 0-12-309770-3

In this lecture, I shall confine my attention to the special problem of free dendritic growth of a pure crystal from its melt. This is the clearest example of emerging pattern selection that I know of. It seems to have no exact analog in hydrodynamics, probably because of the essential role played by crystalline anisotropy, and therefore is particularly interesting from a metallurgical point of view. By the term "free dendrite," I mean an isolated dendritic structure growing into an undercolled melt. This situation is distinct from, for example, cooperative growth of cellular or dendritic arrays during directional solidification, where the existence of sharp pattern-selection mechanisms is experimentally less certain. In the elegant experiments of Glicksman and his collaborators,[6,7] dendrites seeded at the center of undercooled fluids have growth rates, tip radii, and sidebranch spacings which are reproducibly determined only by the undercooling. That is, no matter what the shape of the initial seed or the detailed form of the instability which instigates the dendritic behavior, the leading tip of each dendrite quickly finds a natural speed and shape governed only by steady-state growth conditions and independent of prehistory.

This sharply selective behavior of the dendrite has seemed specially puzzling because the equations of motion for solidification fronts at fixed undercooling appear to have not one, but a whole family, of dendrite-like solutions. These are the so-called "needle-crystals," shape preserving liquid-solid interfaces which are paraboloidal at their tips and whose growth speeds generally decrease with increasing tip radius.[8] The existence of these solutions for non-vanishing capillarity has never been proven rigorously; and there is reason to suspect that the picture that has emerged from various approximate analyses[9,10] will have to be modified. I shall return to this point later in connection with the boundary-layer model. For the moment, the important point is that the standard mathematical statement of the solidification problem seems to admit a multitude of steady, propagating, dendrite-like solutions; that nature selects only one of these solutions; and that understanding nature's selection mechanism is a major theoretical challenge.

In 1977, Müller-Krumbhaar and I[11] suggested a selection principle that has, so far, been successful in explaining a number of relevant experimental results. We had noticed that a needle-crystal solution which is sufficiently slow and, accordingly, flat at its tip must become linearly unstable against deformations in which the tip breaks up into sharper, more rapidly growing protuberances. This, of course, is just the Mullins-Sekerka instability on a slowly curved surface.[2] Our suggestion, rephrased here in language more consistent with current understanding, was that an initially smooth, unstable shape might naturally sharpen until it reaches its slowest stable growth mode. In other words, the natural operating mode of the dendrite is at or near the growth rate where its tip is just marginally stable. This principle of marginal stability, although difficult to use without further poorly controlled approximations, has proved to be consistent with Glicksman's data for succinonitrile over five decades of growth velocity.[7] Perhaps more convincing, it has correctly predicted the initial rise in growth velocity that occurs when small concentrations of impurities are added to the melt.[12-14] The latter effect is one whose sign is not even given correctly by previous steady-state theories.

Although the marginal-stability hypothesis has been used with some success in various applications during the last six years or so, until very recently there has been little progress in understanding it from a deeper theoretical point of view. We still do not know when, if ever, it is correct. It might be an exact result; it might be an approximation accurate for small values of some parameter; or it might simply be wrong. As applied so far, marginal-stability calculations involve only equations of motion which are linearized about steady-state needle-crystals solutions or approximations thereof. Thus, although sidebranching instabilities play an essential role in the theory, one gets from these linear calculations no clue about the amplitudes of the sidebranches or their shapes as they grow outwards and restabilize. Nor can one use the theory as developed so far to study how these more detailed aspects of the dendritic structure depend on various parameters such as undercooling, crystalline anisotropy, or the like.

During the last year, there have been two developments which cause me to be optimistic about understanding some of these fully nonlinear aspects of the dendrite problem. First, we have discovered a class of one-dimensional models which, although not realistically descriptive of dendritic solidification, exhibit nontrivial pattern selection in agreement with the marginal-stability hypothesis. Second, guided by the mathematical structure of the one-dimensional models, we have begun to study a more realistic "boundary-layer" model of solidification. Preliminary analytic and numerical results with this new model look promising. Unfortunately, the deadline for this manuscript is too early for me to be able to report answers to the most interesting questions posed in the last paragraph.

The one-dimensional pattern forming models that realize the marginal-stability mechanism are described by differential equations of the form

$$\frac{\partial f}{\partial t} = \Lambda\left(f\,,\ \frac{\partial^2 f}{\partial x^2}\,,\ \frac{\partial^4 f}{\partial x^4}\right) \tag{1}$$

where $f(x,t)$ might be visualized as the displacement of a solidification front at position x and time t. The quantity Λ on the right-hand side of (1) in general is a nonlinear, algebraic function of its arguments. In order for an equation of this kind to describe nontrivial pattern formation, it is essential that Λ contain at least fourth derivatives of f. Several examples of models of this kind have been discussed in the recent literature.[15,16] One of the simplest is

$$\frac{\partial f}{\partial t} = \left[\varepsilon - \left(\frac{\partial^2}{\partial x^2} + 1\right)^2\right] f - f^3 \tag{2}$$

which was first introduced by Swift and Hohenberg[17] and then studied in more detail by Pomeau and Manneville,[18] primarily in the context of hydrodynamic phenomena. The most thorough of our recent investigations[19,20] have dealt with this model, but we have checked that others have the same behavior.

The quantity ε in Eq. (2) is a control parameter chosen so that the planar state $f = 0$ becomes unstable for $\varepsilon > 0$. Small perturbations of the form $f \propto exp(\omega t + ikx)$ grow or decay according to the law

$$\omega(k) = \varepsilon - (k^2 - 1)^2 \ . \tag{3}$$

For ε in the range $0 < \varepsilon < 1$, there are stable stationary solutions of (2) which are periodic functions of x. At any given ε, there exists not just one such solution but a continuous band of them, say $f_k(x)$, with periodicities varying across some range of values of the fundamental wavenumber k. Thus this system presents a well-posed pattern selection problem: which, if any, of the solutions $f_k(x)$ will be generated by given initial conditions.

Equation (2), unlike most equations of the general form (1), permits a special answer to a pattern-selection question which is not, however, precisely the question posed above. We can write (2) in the form

$$\frac{\partial f}{\partial t} = -\frac{\delta F}{\delta f} \tag{4}$$

where $\delta / \delta f$ denotes a variational derivative and

$$F\{f\} = \int dx \left[\frac{1}{2}\left(\frac{\partial^2 f}{\partial x^2}\right)^2 - \left(\frac{\partial f}{\partial x}\right)^2 + \frac{1}{2}(1-\varepsilon)f^2 + \frac{1}{4}f^4\right]. \tag{5}$$

Then, for all f,

$$\frac{dF}{dt} = -\int dx \left(\frac{\delta F}{\delta f}\right)^2 \leq 0 \ , \tag{6}$$

so that F is a Lyapunov function. Thus any given $f(x)$ must move according to (2) toward a local minimum of F. Each of the stable f_k is such a minimum. If the system is subject to thermal noise (or its equivalent caused by other noisy perturbations), then the most probable state will be the f_k for which $F\{f_k\}$ is the absolute minimum of F. My present bias is to doubt the relevance of such fluctuation-driven selection mechanisms in solidification theory. True thermal fluctuations seem to be many orders of magnitude too small to drive the observed phenomena; and other noise sources such as macroscopic inhomogeneities would not seem to lead to reproducible behavior. Moreover, I do not think that adding stochasticity to the theory is necessary. Purely deterministic, nonlinear, dissipative models seem to contain pattern-selection mechanisms of their own.

The possible relevance of these one-dimensional models to dendritic growth appears in connection with what we have called propagating pattern selection.[19,20] Consider an initially structureless system, $f = 0$, which is "quenched" to some $\varepsilon > 0$ so that it becomes uniformly unstable against small deformations. A perturbation which at first is confined to a small region will grow locally into a well-developed oscillatory pattern, and this pattern will spread throughout the rest of the space. Analytic arguments and extensive numerical experiments indicate that the pattern spreads by propagating at a well-defined velocity, the front of the pattern looking much like the tip of a dendrite which generates an array of sidebranches behind it as it moves. A picture of such a pattern front is shown in Fig. 1; the nodes are stationary relative to the x axis and new oscillations are emerging at the front.

A remarkable fact about this process is that, so long as the starting perturbation is well-localized, both the speed of propagation and the wavelength of the pattern are completely independent of the shape of that perturbation, and the wavelength is *not* the one which minimizes the Lyapunov function F. There are special initial configurations which can produce patterns that propagate at other speeds, with other periodicities; but these configurations have exponentially small, oscillating tails which extend into the otherwise unperturbed, unstable region of the system, and which would be essentially impossible to prepare in a real experiment. The physically accessible, localized initial configurations all produce the same propagating state; thus this is an example of a sharp selection mechanism.

A second remarkable fact is that this selected state is consistent with the marginal-stability principle. The term "stability" is used here in just the same sense as used above in connection with dendritic growth. That is, we look in the frame of reference moving with the front and ask whether an initially localized

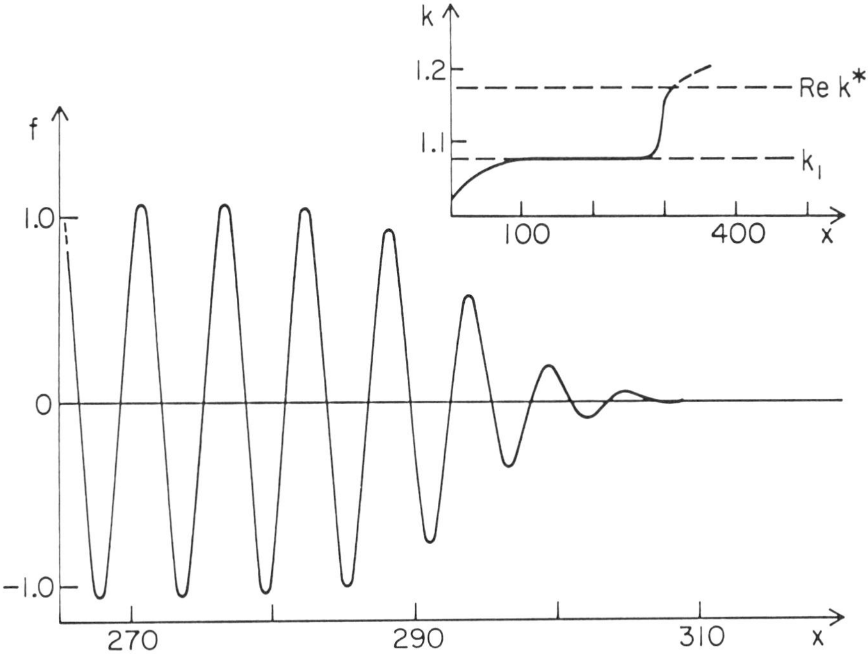

Figure 1

Front portion of a propagating pattern determined by Eq. (2) for $\varepsilon = 0.9$. The oscillatory part of the pattern on the left is stationary in the laboratory frame, and new oscillations arise as the envelope of the pattern moves to the right. Inset: the local wave number k as a function of x for the entire system. The values k_1 and k^* are marginal-stability predictions for the selected wave number in the body of the pattern and the initial wave number which emerges ahead of the front.

perturbation, observed at a fixed point in that frame, will grow or decay. A perturbation which decays is considered stable even if it generates a growing disturbance, like a sidebranch, which moves away from its point of origin near the front. No completely rigorous stability analysis for the propagating solutions of (2) has yet been carried out; and the difficulties are such that it is unlikely this will be done in the near future. However, a systematic and convincing analysis can be made in the limit of small positive ε, and a more speculative procedure gives answers at all ε. The latter procedure, which has been described in more detail in several other publications,[19-21] uses the equations of motion linearized about $f = 0$ to study properties of the leading edge of the pattern front. The procedure can be shown to be equivalent to a partial stability analysis and, as such, should give a propagation speed which is either exact or is a lower bound to the correct value. When applied to Eq. (2), the method predicts propagation speeds which are in excellent agreement with numerical results.

I should like to turn now to the boundary-layer model. Having noticed that a roughly dendritic kind of pattern selection occurs in a deterministic way in certain one-dimensional dynamical systems, it seems natural to ask whether similar behavior might be found in more realistic models. The mathematical side of the strategy for answering this question is most easily discussed for two dimensional solidification problems, but is generalizable in principle to fully three-dimensional situations. The idea is to replace the linear position x by the arc length s measured along the solidification front. The natural replacement for the displacement of the front is its curvature, K. Knowing K as a function of s, it is possible to reconstruct the entire shape of the growing solid. An equation of motion for K must have the form

$$\left(\frac{dK}{dt}\right)_n = -\left(\frac{\partial^2}{\partial s^2} + K^2\right)v_n \tag{7}$$

where v_n is the normal growth rate of the front and $(d/dt)_n$ denotes the rate of change along the normal growth direction. Equation (7) is simply a geometric identity. The content of the theory is determined by the way in which v_n depends on K and other variables of interest.[22]

Clearly (7) is a nonlinear equation for $K(s,t)$ of the same general structure as (1), although we shall see that there are important mathematical differences between the models we shall want to study using (7) and those discussed above. The crucial simplification that allows (7) to be more tractable than the full solidification problem is the choice of $v_n(s,t)$ to be a local function, that is, to depend only on quantities evaluated at the position s and time t so that (7) is a differential rather than an integral equation. This is not quite realistic. The actual motion of a point on a solidification front is determined by the thermal field near that point which, in turn, is determined by the latent heat which has been generated at earlier times at neighboring points. Thus the full solidification problem is nonlocal in both space and time.

The boundary layer model[23-25] incorporates some part of this nonlocality in the dynamics of a new thermal field h which is defined not in the entire two (or three) dimensional space, but only along the liquid-solid interface; in other words, h is a function of s. This function may be interpreted as the heat content per unit length (or area) of a thermal layer in the liquid which contains the latent heat that has been rejected by the advancing solid. By allowing this heat to diffuse along the boundary, we preserve some features of the nonlocal dynamics that seem essential for realistic dendritic behavior. However, physical validity of the model requires that the boundary layer be thin compared to the radius of curvature of the solidification front. In technical terms, the Peclét number must be large. This condition is not satisfied in many of the most interesting experimental situations, for example, dendrites growing at small undercooling. But there are other situations, such as thermal dendrites at large undercooling or chemical dendrites, where the condition is satisfied; and under these circumstances the boundary-layer model is actually a realistic approximation. Even under circumstances where the model is not fully realistic, it may be a useful mathematical model of pattern selection.

The boundary-layer model is constructed as follows. Let u_s be the dimensionless difference in temperature between the liquid-solid interface and the undercooled fluid infinitely far from the solid, measured in units of the ratio of the latent heat to the specific heat. The Gibbs-Thomson condition supplemented by a simple model of interfacial attachment kinetics requires

$$u_s = \Delta - d_0 K - \beta v_n \tag{8}$$

where Δ is the dimensionless undercooling. The capillary length d_0 and the kinetic coefficient β may be functions of the orientation of the crystalline surface at the point where (8) is applied. Next we define a length l, or equivalently the thermal field h, such that l is the effective thickness of the thermal boundary layer and $h = u_s l$ is its heat content per unit length. If we neglect heat flow in the solid and approximate the normal temperature gradient in the fluid by u_s / l, then heat conservation at the surface requires

$$v_n = \frac{D u_s^2}{h} \ , \tag{9}$$

where D is the thermal diffusion constant. Equation (9) is to be used on the right-hand side of (7) to produce an equation of motion for K.

A suitable equation of motion for h can be written in the form

$$\left(\frac{dh}{dt}\right)_n = v_n (1 - u_s) + D \frac{\partial}{\partial s}\left(l \frac{\partial u_s}{\partial s}\right) - h v_n K \ . \tag{10}$$

The first term on the right-hand side is the rate at which latent heat is being added to the boundary layer, and the second term accounts for lateral diffusion of this heat along the surface. The third term is a geometrical correction which can be identified as the origin of the Mullins-Sekerka instability. A surface element of positive curvature (outward bulge) increases in length as it grows, thus thinning the boundary layer, sharpening the thermal gradient, and, via (9), increasing v_n.

The above equations, supplemented by a simple geometric prescription for updating values of s on the expanding boundary, form a complete dynamical system. We have been able to verify analytically that this system accurately reproduces known special solutions of the full solidification problem.[24] For example, the time dependence of a growing circular solid and its associated Mullins-Sekerka instability are reproduced correctly. The only significant discrepancy is that the Δ-dependence of these solutions is incorrect in the limit $\Delta \ll 1$ where the boundary-layer picture is known to be wrong. In the limit of vanishing d_0 and β in (8), the system possesses a family of unstable parabolic needle-crystal solutions as found by Ivantsov.[8] The equations are sufficiently tractable, however, that we can show that these solutions do not survive the inclusion of capillarity. I think it is possible that the same thing happens in the full solidification problem. The presumably exact maximum-velocity calculations of Nash and Glicksman[26] assume the existence of steady-state solutions but do not really prove it.

Some of our preliminary numerical results are shown in Figs. 2 through 4. Figure 2 shows a sequence of solidification fronts computed in the (unphysical) limit $\Delta \ll 1$ and with no crystalline anisotropy, that is, $\beta = 0$ and $d_0 = $ constant in (8). The initial shape was a circle with a small six-fold anisotropy; and six-fold symmetry was enforced at all later stages. Note that we are able to follow the Mullins-Sekerka instability far into the nonlinear regime where the outward bulges flatten and the heat accumulates in deep grooves. This behavior is generally characteristic of unstable solidification fronts observed in experiments; but, so far as I know, this is the first time it has been reproduced theoretically. Also apparent in the figure is that this process does not produce a snowflake. As verified by further computation, the next grooving instability occurs at the

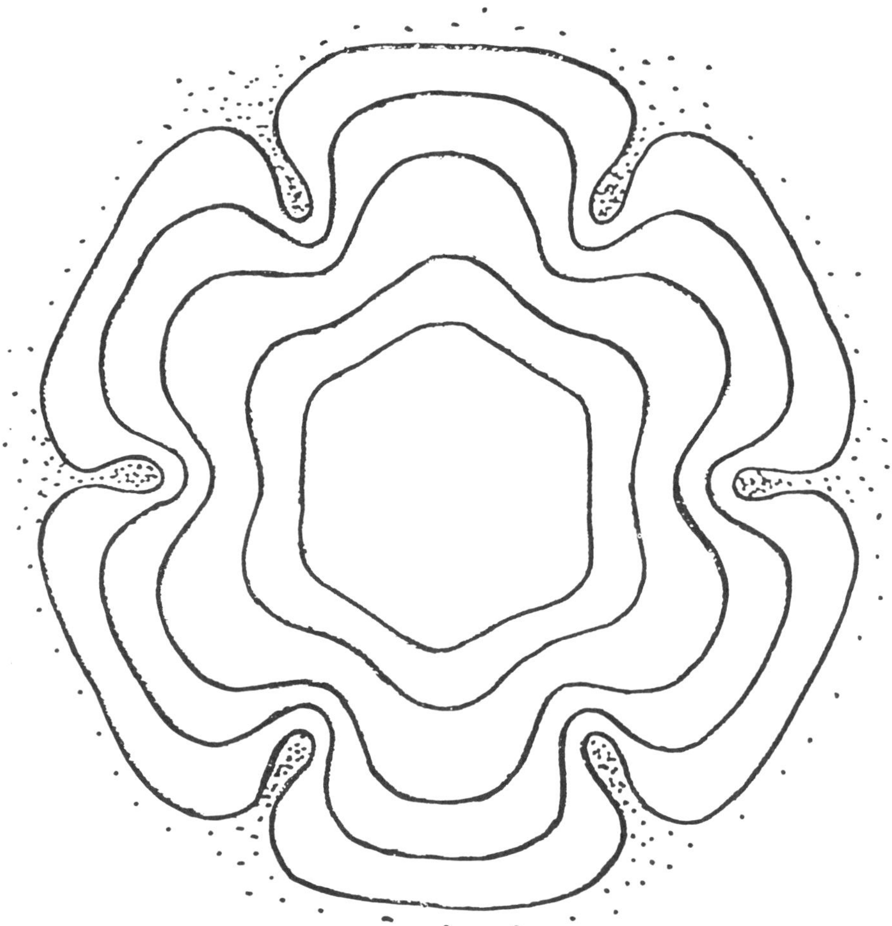

Figure 2

Time-ordered sequence of solidification fronts starting from a circle with a small six-fold anisotropy but with purely isotropic growth kinetics. The dots simulate schematically the accumulation of latent heat in the grooves.

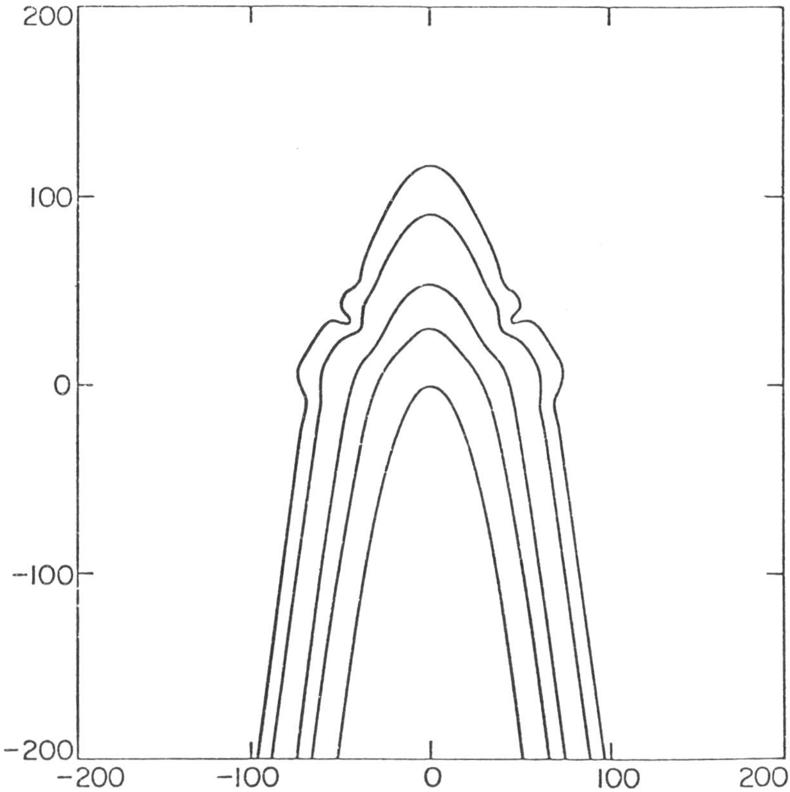

Figure 3

Time-ordered sequence of solidification fronts starting from a parabola and including a four-fold anisotropy in the growth kinetics. Times shown, in units of $d_0^2 D^{-1} \Delta^{-8}$, are 10, 60, 111, 186, and 240. Lengths are in units $d_0 \Delta^{-3}$.

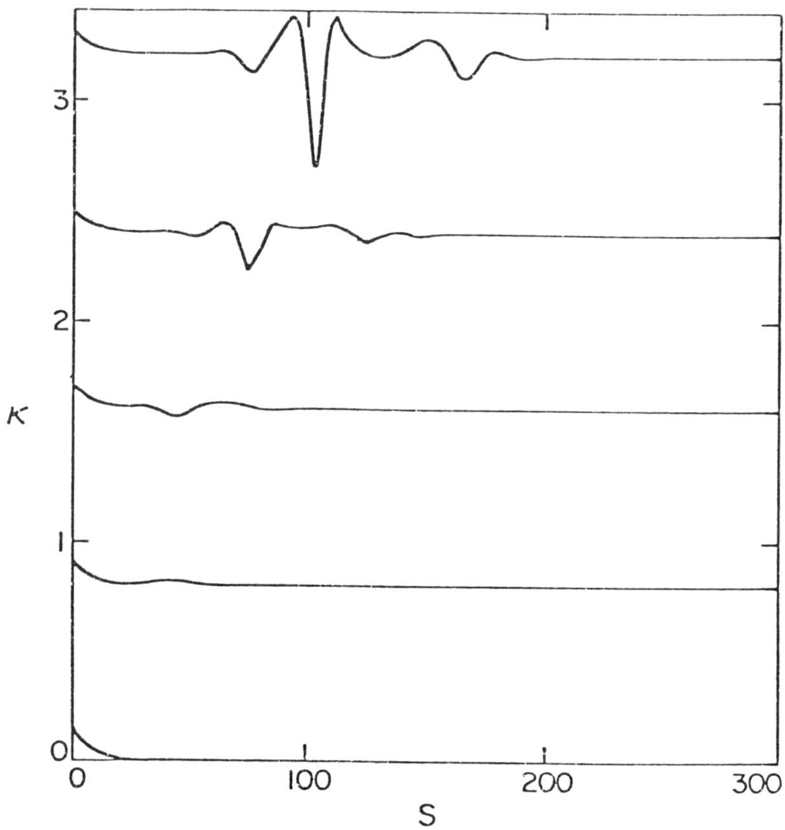

Figure 4

Graphs of curvature K(s,t) corresponding to the solidification fronts shown in Fig. 3. Each successive curve has been shifted up along the K axis by 0.8.

flattened centers of the outward bulges, producing a pair of fingers which, themselves, will later flatten and split. If continued indefinitely, this system would produce an increasingly complex fractal structure which might be of some interest in biology but probably not in crystal growth.

Another feature not shown explicitly in Fig. 2 is that the outward bulges on opposite sides of a groove will, in this model, appear to grow through one another. This is because the boundary-layer model is missing the long-range interactions via the thermal field that would, in reality, prevent two separate pieces of the solidification front from growing together. For purposes of studying selection mechanisms in dendritic tips, this does not seem to us to be a serious drawback. (A way to avoid this difficulty has been suggested by Müller-Krumbhaar.[27]) A potentially more interesting but highly speculative feature of this kind of "non-crystalline" process is the way in which the six-fold symmetry seems to be unstable if not rigorously enforced computationally. We suspect that this system, if left to evolve after a very small symmetry breaking perturbation, would produce an intrinsically chaotic picture more reminiscent of seaweed than any regular pattern.

The missing ingredient in Fig. 2 for producing regular dendritic behavior seems to be crystalline anisotropy. This, too, is fairly clear from the figure. If we choose d_0 or β in (8) to be functions of orientation which enhance growth along the hexagonal directions and suppress it in between, then the forward bulges may remain sharp, and the flat regions which will undergo the next grooving instabilities may appear at the sides of these embryonic dendrites. This kind of behavior is shown in Fig. 3. In this sequence, $\Delta = 0.9$ and we have chosen a four-fold kinetic anisotropy with $\beta = 0.1(1 - \cos 4\Theta)$, Θ being the angular orientation of the surface. The initial shape is a parabola. The figure is reminiscent of Glicksman's pictures of succinonitrile dendrites. The tip is smooth, very nearly parabolic, and has settled quickly to what appears to be a steady-state velocity and tip curvature. The sidebranching deformation is very weak near the tip, and rises abruptly as grooving sets in at some distance back. The corresponding graphs of $K(s,t)$ are shown in Fig. 4. Here one sees in much more detail than in real space how the simple deformations which arise near the tip at $s = 0$ grow and become more structured as they move back along the dendrite.

I shall close by listing a few questions:

If the sequence shown in Fig. 3 is indeed approaching a steady-state dendritic growth velocity, does this velocity correspond to the point of marginal stability of the tip? Is this tip behavior produced uniquely, independent of starting configuration?

Is it necessary to have a finite strength of anisotropy in order to produce dendritic behavior? Or is any nonzero anisotropy sufficient?

How do the sidebranch spacing, the amplitude of the initial sidebranching deformation, and the distance from the tip at which the first large sidebranches appear depend on undercooling Δ and the strength of the anisotropy? (Other parameters such as D and d_0 can be scaled out of the theory by redefining length and time scales.)

We do not yet have answers to these questions; but each of these questions now seems answerable via the boundary-layer model.

ACKNOWLEDGMENTS

This lecture is based largely on ongoing research being carried out in collaboration with E. Ben-Jacob, G. Dee, N. Goldenfeld, and G. Schön. The projects are supported by the Department of Energy under Contract No. DE-AM03-765F00034 and by the National Science Foundation under Grant No. PHY77-27084 with supplemental funds from the National Aeronautics and Space Administration.

REFERENCES

1. J. S. Langer, Materials Science and Engineering 65, pp. 37-44 (1984).
2. W.W. Mullins and R.F. Sekerka, J. Appl. Phys. **34**, 323 (1963).
3. W.W. Mullins and R.F. Sekerka, J. Appl. Phys. **35**, 444 (1964).
4. J.S. Langer, Rev. Mod. Phys. **52**, 1 (1980).
5. S. Chandrasekhar, *Hydrodynamic and Hydromagnetic Stability* (Oxford University Press, London, 1961).
6. M.E. Glicksman, R.J. Shaefer, and J.D. Ayers, Metall. Trans. A **7**, 1747 (1976).
7. S.C. Huang and M.E. Glicksman, Acta Metall. **29**, 701 (1981); **29**, 717 (1981).
8. G.P. Ivantsov, Dokl. Akad. Nauk SSSR **58**, 567 (1947).
9. D.E. Temkin, Dokl. Akad. Nauk SSSR **132**, 1307 (19).
10. R.F. Sekerka, R.G. Seidensticker, D.R. Hamilton, and J.D. Harrison, "Investigation of Desalination by Freezing," Westinghouse Research Laboratory Report (1967).
11. J.S. Langer and H. Müller-Krumbhaar, Acta Metall. **26**, 1681; 1689; 1697 (1978).
12. J.S. Langer, Physicochemical Hydrodynamics **1**, 41 (1980).
13. M.A. Chopra, "Influence of Diffusion and Convective Transport on Dendritic Growth in Dilute Alloys," Ph.D. Thesis, Rensselaer Poytechnic Institute (1983).
14. A. Karma and J.S. Langer, to be published.
15. G. Sivashinsky, Ann. Rev. Fluid Mech. **15**, 179 (1983).
16. J.S. Langer and H. Müller-Krumbhaar, Phys. Rev. A **27**,499
17. J. Swift and P.C. Hohenberg, Phys. Rev. A **15**, 319 (1977).
18. Y. Pomeau and P. Manneville, J. Phys. Lett. **40**, 609 (1979).
19. G. Dee and J.S. Langer, Phys. Rev. Lett. **50**, 383 (1983).
20. E. Ben-Jacob, H. Brand, G. Dee, L. Kramer, and J.S. Langer, to be published.

21. J.S. Langer, Proceedings of the AIME Symposium on "Establishment of Microstructural Spacing during Dendritic and Cooperative Growth," March, 1983; Metall. Trans., in press.

22. "String" models in which v_n depends only on K and $\partial^2 K/\partial s^2$ have been studied recently by R. Brower, D. Kessler, J. Koplik, and H. Levine, Phys. Rev. Lett. **51**, 1111 (1983).

23. E. Ben-Jacob, N. Goldenfeld, J.S. Langer, and G. Schön, Phys. Rev. Lett., in press.

24. E. Ben-Jacob, N. Goldenfeld, J.S. Langer, and G. Schön, Phys. Rev. A, in press.

25. The boundary-layer model has its roots in some of the earliest attempts to solve solidification problems. For example, see C. Zener, J. Appl. Phys. **20**, 950 (1949). For an early example of the use of curvature as a dynamical variable in metallurgical free boundary problems, see W.W. Mullins, J. Appl. Phys. **27**, 900 (1956); **28**, 333 (1957).

26. G.E. Nash and M.E. Glicksman, Acta Metall. **22**, 1283 (1974).

27. H. Müller-Krumbhaar, Proceedings of the NATO Workshop on Chemical Instabilities, Austin, Texas (1983).

Institute for Theoretical Physics
University of California
Santa Barbara, California 93106

MORPHOLOGICAL INSTABILITIES DURING PHASE TRANSFORMATIONS

R. F. Sekerka

1. INTRODUCTION

This paper deals with a free boundary problem that is a generalization of the classical Stefan problem. The problem arises in the context of phase transformations and pertains to the shape of the boundary that separates a growing phase from a nutrient phase under conditions for which the growth is controlled by diffusive transport of heat and/or solute. For simple geometries and far-field conditions, it is often possible to find exact solutions for phase boundaries with simple shapes, (e.g., spheres, cylinders and planes) but under a variety of growth conditions, such simple shapes are unstable. These so-called morphological instabilities have been the subject of numerous review articles. [1-8]

In the present paper, we use the solidification of a pure substance as a prototype; we recast the problem, by using dimensionless variables, in a form that is more suitable for mathematical analysis.

2. SPHERE PROBLEM

We consider the solidification of a pure solid (subscript S) occupying a closed domain D, from an infinite bath of pure convectionless liquid (subscript L) having the same density as the solid. The temperatures T_L and T_S obey the equations

ISBN 0-12-309770-3

$$\nabla^2 T_L = (1/\alpha_L)\ \partial T_L/\partial t \tag{1}$$

in the domain D

and $$\nabla^2 T_S = (1/\alpha_S)\ \partial T_S/\partial t \tag{2}$$

outside the domain D, where α_L and α_S are constant thermal diffusivities, ∇^2 is the Laplacian operator in dimensional coordinates, and t is the time. On the boundary, ∂D, we have two conditions, (1) the conservation of energy

$$(-k_L\ \nabla T_L + k_S\ \nabla T_S) \cdot \hat{n} = \left[L_o + \rho_L (C_L - C_S)(T_I - T_M)\right] U \tag{3}$$

and (2) the condition of local thermodynamic equilibrium

$$T_I = T_M - T_M \Gamma K \tag{4}$$

where k_L, C_L and k_S, C_S are thermal conductivities and specific heats of the liquid and solid, respectively, ρ_L is the density of either phase, $\hat{n}$ is a unit normal to ∂D pointing into the liquid, L_o is the latent heat of fusion per unit volume at temperature T_M, U is the local normal growth speed, T_I is the temperature on the boundary (I because this boundary is often called the solid-liquid interface), T_M is the absolute melting point of the bulk solid, Γ is a capillarity length (usually of atomic dimensions and proportional to the solid-liquid surface tension) and K is the mean curvature of the boundary (sign convention such that $K = 2/R$ for a sphere of solid of radius R). We include the term in Eq(3) that contains specific heats to avoid inconsistencies that can arise when $C_L \neq C_S$ and also to parallel the usual treatment of the related problem in which growth occurs because of isothermal diffusion. We assume that T_L tends to a bath temperature T_∞ (where $T_\infty < T_M$) as $|\vec{r}| \to \infty$ and defer discussion of the initial shape and initial temperature conditions until later.

For $\Gamma = 0$, the above problem is the three-dimensional classical Stefan problem. For $\Gamma \neq 0$, we shall, therefore, refer to it as the modified Stefan problem and, as we shall see later, this modification is crucial insofar as morphological stability is concerned.

2.1 Dimensionless Formulation

We introduce dimensionless variables by scaling all lengths with a constant length L, time by a constant time T, define dimensionless temperatures

$$u_{S,L} = (T_{S,L} - T_\infty)/(T_M - T_\infty), \tag{5}$$

a dimensionless growth speed

$$\tilde{U} = UT/L, \tag{6}$$

and a dimensionless curvature

$$\tilde{K} = LK. \tag{7}$$

Then Eqns. (1-4) become:

$$\Delta u_L = (L^2/\alpha_L T)(\partial u_L/\partial\tau) \tag{8}$$

$$\Delta u_S = (L^2/\alpha_S T)(\partial u_S/\partial\tau) \tag{9}$$

$$- (\partial u_L/\partial\nu) + q(\partial u_S/\partial\nu) = (L^2/\alpha_L TS)\left[1 + S(1 - \tfrac{q}{\rho})(u_I - 1)\right]\tilde{U} \tag{10}$$

$$u_I = 1 - \left[(T_M\Gamma)/(T_M - T_\infty)L\right]\tilde{K} \tag{11}$$

where Δ is the dimensionless Laplacian operator, $\tau = t/T$ is dimensionless time, ν is dimensionless distance along the interface normal, $q = k_S/k_L$, $\rho = \alpha_S/\alpha_L$, and the parameter

$$S = \left[T_M - T_\infty\right]/\left[L_o/(\rho_L C_L)\right] \tag{12}$$

where the relationship $\alpha_L = k_L/(\rho_L C_L)$ has been used. The parameter S plays the role of a dimensionless undercooling that drives the interface motion, $\tilde{U}$; furthermore, the domain of physical interest is $S << 1$ because larger values will generally result in violation of Eq.(4), the condition of local equilibrium at the moving interface. Thus, we see that the time scales in Eqs.(8) and (9) are different from that in Eq. (10) and we choose to focus attention on the problem of interface motion by taking

$$L^2/(T\alpha_L S) = 1 \tag{13}$$

and then choosing the length scale (see Eq.(11) to satisfy

$$\left[T_M \Gamma / (T_M - T_\infty) L\right] = 1/2. \tag{14}$$

The factor of 1/2 on the righthand side of Eq.(14) makes L equal to the nucleation radius R* (which stems from Eq.(4) by setting $T_I = T_\infty$ and $K = 2/R^*$). Thus, Eqs. (8-12) become:

$$\Delta u_L = S \partial u_L / \partial \tau \tag{15}$$

$$\Delta u_S = (S/\rho) \partial u_S / \partial \tau \tag{16}$$

$$-(\partial u_L / \partial \nu) + q(\partial u_S / \partial \nu) = \tilde{U}\left[1 + S(1-\tfrac{q}{\rho})(u_I - 1)\right] \tag{17}$$

$$u_I = 1 - (1/2)\,\tilde{K} \tag{18}$$

Eqs. (15-18), subject to the condition $u \to o$ as $|\vec{r}| \to \infty$ and a suitable set of initial conditions, serve to define the problem.

2.2 Quasi-steady State Approximation

The fact that S is small, however, suggests that the transients associated with Eqs. (15) and (16) may not be important on a time scale over which significant interface motion can occur. Thus, we may approximate Eqs. (15) and (16) by the Laplace equations

$$\Delta u_L = \Delta u_S = 0 \tag{19}$$

which is known as the quasi-steady state approximation [9]. Within this approximation, a simple spherically symmetric solution may be found in which the solid-liquid interface is a sphere of dimensionless radius $\tilde{R}$,

$$u_S = 1 - (1/\tilde{R}), \tag{20}$$

$$u_L = \left[1 - (1/\tilde{R})\right] (\tilde{R}/\tilde{r}) \tag{21}$$

and

$$d\tilde{R}/d\tau = \left[1 - (1/\tilde{R})\right] (1/\tilde{R}) \tag{22}$$

where $\tilde{r}$ is the dimensionless spherical radius. The nucleation radius, at which $\tilde{R} = 1$, is a stationary solution to Eq.(22); a solution for a growing sphere may be found by integrating from $\tilde{R} = 1 + \varepsilon$ at $\tau = o$ to obtain

$$\tau = -\ln\varepsilon - \varepsilon - \varepsilon^2/2 + \ln(\tilde{R}-1) + \tilde{R} + \tilde{R}^2/2. \tag{23}$$

The effect of finite ε is simply to shift the origin of time by a sort of residual time needed for true growth to begin. We see from Eq. (22) that a maximum growth rate occurs at $\tilde{R} = 2$. For large $\tilde{R}$, the last term on the righthand side of Eq. (23) eventually dominates and we have approximately

$$\tilde{R} \sim (2\tau)^{1/2} \tag{24}$$

which is a solution of the classical Stefan type.

The question of morphological stability now arises with respect to a sphere growing according to Eq. (23). This question was answered within the quasi-steady state approximation by Mullins and Sekerka [9] who studied the growth of a sphere perturbed by a spherical harmonic of small amplitude $\tilde{\delta}$:

$$\tilde{r} = \tilde{R} + \tilde{\delta} Y_{\ell m}(\theta, \phi). \tag{25}$$

Details will not be reproduced here but the following conclusions are noteworthy:

(1) The spherical solution is unstable with respect to the growth of perturbations of harmonic index ℓ whenever $\ell \geqslant 2$ and

$$\tilde{R} > (1/2)\left[(\ell+1)(\ell+2) + 2 + q\ell(\ell+1)\right] \tag{26}$$

and (2) The maximum perturbation growth rate occurs (for large ℓ, treated as a continuous variable) at a dimensionless wavelength.

$$\Lambda = \frac{2\pi\tilde{R}}{\ell} \sim 2\pi\left[3(1+q)/2\right]^{1/2}. \tag{27}$$

Thus, instability first occurs toward an ellipsoidal shape ($\ell=2$) at $\tilde{R} = 7 + 3q$ and toward other shapes at larger values of $\tilde{R}$.

2.3 Time-Dependent Case

Treatment of the fully time-dependent case (Eqs. (15-18)) is much more difficult.

For the classical Stefan case for which the term in Eq. (18) containing K is missing, there is a well-known spherically symmetric solution [10] that corresponds to the initial conditions $u_S = u_L = 0$ at $\tau = 0$; the temperature field u_L may be expressed in terms of the variable $\tilde{r}^2/\tau$ and $\tilde{R}$ increases with time in strict proportion to $\tau^{1/2}$. The stability of this solution has not been analyzed, but it is suspected to be unstable because the capillarity effect (non-zero Γ in Eq. (4), and the term in $\tilde{K}$ in Eq. (18)) is known to be the stabilizing influence within the quasi-steady state approximation.

If the full Eq. (18) is used, spherically symmetric solutions that correspond to various initial conditions may, in principle, be found and tested for stability.

For initial conditions, $u_S = u_L = 0$ and with the further simplifications $\rho = q = 1$, numerical solutions were obtained by Schaefer and Glicksman [11] ; these solutions exhibit a maximum growth rate and a behavior at large R similar to that described by Eq. (23) but they have not been tested for stability.

Krukowski and Turski [12] have exhibited a particular solution in which, surprisingly, $\tilde{R} = \beta\tau^{1/2}$, even though the temperature of the solid-liquid interface is increasing with time according to Eq. (18). Their solution was derived for the case of alloy solidification but it may be specialized to the case of a pure substance. The temperature fields depend not only on $\tilde{r}/\tau^{1/2}$ but also on $\tilde{r}$ explicitly and, surprisingly again, the value of β for their solution is a solution to the equation.

$$\beta\left[1 - \eta\pi^{1/2}\exp(\eta^2)\,\mathrm{erfc}(\eta)\right] = 2, \tag{28}$$

where $\eta = \beta S^{1/2}/2$, which is exactly the same as for the classical Stefan case. The initial conditions for which this solution holds, however, are rather peculiar. For instance, the temperature field in the solid at all $\tau > 0$ is given by

$$u_S(\tilde{r},\tau) = 1 - \frac{1}{\tilde{r}}\frac{\mathrm{erf}(\eta_S)}{\mathrm{erf}(\eta_{SO})} \tag{29}$$

where $\eta_S = (\tilde{r}/2)(S/\rho\tau)^{1/2}$ and $\eta_{SO} = (\beta/2)(S/\rho)^{1/2}$. At some initial time, say τ_o, the solid would have to be in a particular nonisothermal state in agreement with Eq. (29). Similarly, the fluid would have need to have a quite special initial temperature profile. Although the initial conditions needed to obtain this solution may not be realistic, the Kruskowski-Turski solution demonstrates clearly the sensitivity of this problem to the initial conditions. Moreover, since we expect instabilities to occur at a radius given at least approximately by Eq. (26), we only expect stable spherical growth at small radii at which both capillarity and initial conditions have large effects. Clearly, this situation demands a very delicate analysis.

Some progress in treating the transient case has been made by Wey, Gautesen and Estrin [13] for a one-sided model (which relates to the problem of precipitation from solution) that may be represented by Eq. (15), Eq. (17) with q = 0, and Eq. (18). They obtain an approximate transient solution to the spherical growth problem which they proceed to analyze for stability, obtaining a stability criterion of the form

$$2\phi(\ell-1-\phi)(\tilde{R}-1) - S\left[(1-\tilde{R}^{-1})/(1-S\tilde{R}^{-1})\right] + \{(\ell+2)(\ell^2-1) + \phi\left[2-(\ell+2)(\ell-1)\right]\} < 0 \qquad (30)$$

where, in our notation,

$$\phi = S\,\tilde{R}\,d\tilde{R}/d\tau. \qquad (31)$$

For small S, Eq. (22) is valid and $\phi = S(1-\tilde{R}^{-1})$; then to lowest order in S, Eq. (30) yields Eq. (26) with q = 0. As S increases, Wey et al find that the stability of each harmonic decreases slightly and then increases abruptly and significantly at some critical value of S. The $\ell=2$ harmonic, for instance, experiences this abrupt stabilization at $S = 0.344$ at which they show that a sphere grows at the same rate as a paraboloid of revolution with a tip radius equal to the radius of the sphere. This tendency to restabilize for large driving forces is worthy of further explanation and interpretation.

3. PLANE PROBLEM

A related problem arises in the crystal growth situation in which, instead of free growth into an undercooled melt, solidification is caused by heat sources and sinks that travel at constant velocity V. For very long samples, the goal is to achieve steady state unidirectional solidification in which a planar interface separates solid and liquid. In a frame of reference that moves along the z axis with velocity V with respect to the material, Eqs. (1) and (2) are replaced by

$$\nabla^2 T_L + (V/\alpha_L)(\partial T/\partial z) = (1/\alpha_L)(\partial T_L/\partial t) \tag{32}$$

and

$$\nabla^2 T_S + (V/\alpha_S)(\partial T/\partial z) = (1/\alpha_S)(\partial T_S/\partial t) \tag{33}$$

and Eqs. (3) and (4) still apply.

3.1 Dimensionless Formulation

As in section (2.1), we scale all lengths with L and times with T but define a dimensionless temperature

$$\Theta_{S,L} = \frac{T_{S,L} - T_M}{(L_o/\rho_L C_L)} \tag{34}$$

Then Eqs. (32), (33), (3) and (4) become

$$\Delta\Theta_L + (VL/\alpha_L) = (L^2/\alpha_L T)\ \partial\Theta_L/\partial\tau \tag{35}$$

$$\Delta\Theta_S + (VL/\rho\alpha_L) = (L^2/\rho\alpha_L T)\ \partial\Theta_S/\partial\tau \tag{36}$$

$$\partial\Theta_L/\partial\nu - q\ \partial\Theta_S/\partial\nu = (L^2/\alpha_L T)\left[1 + (1-\tfrac{q}{\rho})\Theta_I\right]\tilde{\tilde{U}} \tag{37}$$

$$\Theta_I = -(T_M \Gamma \rho_L C_L/L L_o)\ \tilde{\tilde{K}} \tag{38}$$

where $\tilde{z} = z/L$, $\tau = t/T$, $\tilde{\tilde{U}} = UT/L$ and $\tilde{\tilde{K}} = KL$.

In place of Eqs. (13) and (14), we choose $L^2/\alpha_L T = 1$, $VL/\alpha_L = 1$ and define

$$\varepsilon = (T_M \Gamma \rho_L C_L V/\alpha_L L_o). \tag{39}$$

Then Eqs. (35)-(38) become

$$\Delta\Theta_L + \partial\Theta_L/\partial\tilde{z} = \partial\Theta_L/\partial\tau \tag{40}$$

$$\Delta\Theta_S + (1/\rho)\partial\Theta_S/\partial\tilde{z} = (1/\rho)\partial\Theta_S/\partial\tau \tag{41}$$

$$-\partial\Theta_L/\partial\nu + q\partial\Theta_S/\partial\nu = \left[1 + (1-\tfrac{q}{\rho})\ \Theta_I\right]\tilde{\tilde{U}} \tag{42}$$

$$\Theta_I = -\ \varepsilon\ \tilde{\tilde{K}}. \tag{43}$$

A similar scaling has been used by Langer [7] but differs

from that used in our section (2.1) where we chose a scaling that related easily to the quasi-steady state approximation. In the present case, a scaling that leads to that approximation is only possible for certain values of the boundary conditions that remain to be specified.

3.2 Unperturbed Solution

Equations (40)-(43) admit a steady state solution (in the moving frame of reference) for which the solid-liquid interface is located in the plane $\tilde{z} = 0$. This so-called unperturbed solution, that we shall later analyze for stability, is just

$$\Theta_L^o = g_L \left[1 - \exp(-\tilde{z})\right] \tag{44}$$

$$\Theta_S^o = \rho g_S \left[1 - \exp(-\tilde{z}/\rho)\right] \tag{45}$$

where g_L and g_S are dimensionless temperature gradients in the liquid and solid, respectively (they are related to the dimensional gradients G_L and G_S by $g_L = G_L k_L/(L_o V)$ and $g_S = G_S k_L/(L_o V)$). Indeed, at $\tilde{z} = 0$, $\partial \Theta_L^o/\partial \tilde{z} = g_L$ and $\partial \Theta_S^o/\partial \tilde{z} = g_S$. We note that Θ_L^o approaches a finite value as $\tilde{z} \to \infty$ but that Θ_S^o becomes infinite as $\tilde{z} \to -\infty$. We, therefore, understand Eq. (45) to be an asymptotic solution for $\tilde{z} < 0$ which, in reality, represents approximately a physically meaningful boundary condition at a finite value of $\tilde{z}$; such a boundary condition could, for example, be imposed by a traveling furnace of the type used in crystal growth. For a planar interface, $\Theta_I = 0$ so substitution of Eqs. (44) and (45) into Eq. (42) yields

$$-g_L + q\, g_S = 1 \tag{46}$$

Langer [7] has treated the case $g_S = 0$ (i.e., $\Theta_S^o = 0$) which requires $g_L = -1$ which corresponds to a unique undercooling at $\tilde{z} \to \infty$; other planar steady states exist, however, for $g_S \neq 0$.

3.3 Perturbed Solution

We next treat the case of a perturbed planar interface of the form

$$\tilde{z}_I = \tilde{\delta} \exp(\sigma\tau) \cos(k\tilde{x}), \tag{47}$$

where $\tilde{\delta}$ is a small amplitude, k is the wavenumber of the perturbation and σ is a parameter to be determined subsequently. To first order in $\tilde{\delta}$, the solutions to Eqs. (40), (41) subject to Eq. (43) may be written

$$\Theta_L = \Theta_L^o - (g_L + \varepsilon k^2) \exp(-p_L\tilde{z})\ \tilde{z}_I \tag{48}$$

$$\Theta_S = \Theta_S^o - (g_S + \varepsilon k^2) \exp(p_S\tilde{z})\ \tilde{z}_I \tag{49}$$

where

$$p_L = \frac{1}{2} + (\frac{1}{4} + k^2 + \sigma)^{1/2} \tag{50a}$$

and

$$p_S = -\frac{1}{2\rho} + (\frac{1}{4\rho^2} + k^2 + \frac{\sigma}{\rho})^{1/2} . \tag{50b}$$

Substitution into Eq. (42) then yields the following equation from which σ may be determined:

$$\sigma = -g_L(p_L-1) - q\ g_S(p_S + \frac{1}{\rho}) - \varepsilon k^2\left[(p_L-1) + q(p_S + \frac{1}{\rho})\right] . \tag{51}$$

Under conditions for which k^2 dominates the square roots in Eqs. (50), Eq. (51) becomes approximately

$$\sigma \simeq -(g_L + q\ g_S)\ k - \varepsilon k^3 . \tag{52}$$

This approximation is equivalent to the replacement of Eqs. (41) and (42) by $\Delta\,\Theta_L = 0$ and $\Delta\,\Theta_S = 0$ which combines the quasi-steady state approximation of section 2.2 with the neglect of the $\partial/\partial\tilde{z}$ terms that stem from transformation into a moving frame of reference. According to Eq. (52), σ will be negative, and, therefore, the planar interface will be stable, if

$$g_L + q\ g_S > 0 . \tag{53}$$

In terms of dimensional quantities, Eq. (53) is equivalent to $G^* > 0$ where $G^* = (k_S G_S + k_L G_L)/(k_S + k_L)$; stability is, therefore, determined by a conductivity weighted average of the temperature gradients in the solid liquid. Under conditions for which $g_L + q\, g_S < 0$, there is instability for perturbations of sufficiently long wavelength; i.e., for k less than the critical value (marginal stability)

$$k_{crit} = \left[\frac{-(g_L + q\, g_S)}{\varepsilon(1 + q)}\right]^{1/2} \quad . \tag{54}$$

The dimensional wavelength that corresponds to this value of k is just

$$\lambda = \frac{2\pi}{k}\frac{\alpha_S}{V} = 2\pi\left(\frac{T_M\Gamma}{-G^*}\right)^{1/2} \tag{55}$$

which is seen to be a geometric mean of a capillary length, Γ, and a thermal length, $T_M/(-G^*)$.

Under conditions for which Eq. (52) is not valid, the analysis is more complicated because of the different square roots that appear in p_L and p_S. A somewhat more tractable case occurs for $q = \rho = 1$ for which Eq. (51) becomes

$$\sigma = 2\left(\frac{1}{4} + k^2 + \sigma\right)^{1/2}\left[-\varepsilon k^2 - \frac{g_S + g_L}{2}\right] - \frac{1}{2} \quad . \tag{56}$$

Analysis of Eq. (56) subject to the condition that Re $(\frac{1}{4} + k^2 + \sigma)^{1/2} > 0$, which is needed to insure proper behavior of the perturbed part of Θ_S as $z \to -\infty$, leads to a number of possible cases depending on the values of the parameters ε, k^2 and $(g_S + g_L)/2$. For some values of the parameters, there are no allowed values of σ. In other cases, there are a pair of conjugate complex values of σ but always with negative real parts. Positive values of σ occur for $(g_S + g_L)/2$ less than

$$\left(\frac{g_S + g_L}{2}\right)_{crit} = -\frac{3}{4}\,\varepsilon^{1/3} + \frac{1}{4}\,\varepsilon \tag{57}$$

which, for small ε, agrees approximately with Eq. (53) for $q = 1$. Thus, the planar interface is stable for a positive average temperature gradient but becomes unstable once a small critical negative value is exceeded.

4. DISCUSSION

The simple examples discussed in sections 2 and 3 are intended to be illustrative of a transient problem and a steady-state problem. The instabilities that obtain in both cases are, however, believed to have the same fundamental origin, namely the bunching or rarefaction of isotherms near a relative protrusion or a depression in an otherwise smooth shape. This phenomenon has been referred to [9] as the "point effect of diffusion" and seems to be associated with properties of the Laplacian operator that occurs in Eqs. (15), (16), (35) and (36). It would be interesting to understand the instability better in terms of fundamental properties of the Laplacian; perhaps it is possible to define a class of free boundary problems that share the same fundamental instability phenomenon.

As we have seen in section 2, the transient problem has the additional complication that initial conditions must be specified and questions of instability must be framed with respect to those initial conditions. Treatment of the problem within the quasi-steady state approximation allows us to deal with a restricted class of temperature fields in which the initial conditions only enter through the initial particle size (see Eq. (23)). Initial conditions that are not spherically symmetric would certainly be expected to cause deviations from a spherical shape, but these deviations might disappear in time. How can the transient problem be formulated in an unambiguous and meaningful manner with respect to arbitrary initial conditions?

For both the transient and the steady-state problem, there is the question of what happens after the initial shapes become unstable. Under what conditions do other growth forms evolve that propagate without change of shape? The answers to this and related questions depends on a non-linear analysis of the problem; such analysis generally requires the use of numerical techniques [13], but expansion techniques [14,15] may be used for shapes that are not too different from the simple shapes that are unstable. Langer [7,16,17] has advanced the idea that certain shapes propagate under conditions of marginal stability, i.e., under growth conditions such that a small change of growth parameters in one direction will produce instability; presumably, changes in the opposite direction are compensated by non-linear effects that are not completely understood. Apparently needle-shaped crystals (approximately paraboloids of revolution) propagate into undercooled melts with speeds and tip radii that agree with the marginal stability hypothesis [18-20]. Such needle-shaped crystals are believed to be related to the primary stalks of dendrites, the tree-like forms that are observed experimentally to solidify chaotically in undercooled melts.

An important variation of the present problem arises for the case of solidification of an alloy rather than a pure substance. In this case, a solute is present in the melt and this solute becomes redistributed on solidification. A diffusion equation governs the transport of solute in the melt, conservation of solute is required at the solid-liquid interface, and the interface temperature (see Eq. (38)) is shifted by a term that may be taken to be proportional to solute concentration for dilute alloys. New scales for length and time appear and analysis [21] shows that concentration gradients play a strong destabilizing role, according to which they can offset the stabilizing influence of positive average temperature gradients. Of special interest in this case is the analysis of non-planar steady-state shapes [14,15,22] that are supposed to be models of cellular solidification fronts that are observed experimentally.

Another interesting complication occurs whenever one allows for fluid convection in the nutrient phase. The equations for fluid flow must then be solved simultaneously with those for the transport of heat and/or solute with concomitant modification of the transport equations to include convective terms. Analysis of this problem has led to traveling waves and enhanced stability in the case of forced convection

[23-24] and convection cells of the Bénard type in the case of natural convection [25]. The role of fluid flow in morphological stability has been the subject of a recent review [26].

Finally, it should be noted that a further modification can occur whenever the motion of the phase boundary is so rapid that local equilibrium is not a good approximation at the solid-liquid interface. In this case, the interface temperature departs from the value specified by Eq. (4) by an amount that depends on local growth speed and that can also be a function of local crystallographic orientation. In this case, directional effects [27] and oscillatory instabilities [28] can occur.

In summary, the modified Stefan problem in more than one dimension is rich in instability phenomena, a primitive case arising whenever the melting temperature depends on interface curvature and more complicated cases arising whenever solute, fluid convection and non-equilibrium interfacial effects are considered. This is undoubtedly fertile ground for mathematical analysis.

REFERENCES

1. Sekerka, R. F., J. Crystal Growth, 3, 4 (1968) 71-81.
2. Chernov, A. A., Soviet Physics Crystallography 16 (1972) 734-753.
3. Sekerka, R. F., in: Crystal Growth: An Introduction, P. Hartman, ed. (North-Holland, Amsterdam, 1973) pp. 403-443.
4. Delves, R. T., in: Crystal Growth, B. R. Pamplin, ed. (Pergamon, Oxford, 1974), pp. 40-103.
5. Chernov, A. A., J. Crystal Growth 24/25 (1974) 11-31.
6. Wollkind, D. J., in: Preparation of Properties of Solid State Materials, W. R. Wilcox, ed. (Dekker, New York, 1979), pp. 111-191.
7. Langer, J. S., Rev. Mod. Phys. 52 (1980) 1-28.
8. Sekerka, R. F., in: Encyclopedia of Materials Science and Engineering (Pergamon, Oxford), to be published.
9. Mullins, W. W. and R. F. Sekerka, J. Appl. Phys. 34 (1963) 323-329.

10. Carslaw, H. S. and J. C. Jreger, Conduction of Heat in Solids, Oxford University Press, London (1959), second edition, p. 294.
11. Schaefer, R. J. and M. E. Glicksman, J. Crystal Growth 5 (1969) 44.
12. Krukowski, St. and L. A. Turski, J. Crystal Growth 58 (1982) p. 631.
13. Smith, J. B., J. Computational Physics 39 (1981) 112.
14. Wollkind, D. J. and L. A. Segel, Phil. Trans. Roy. Soc. London 268 (1970) 351.
15. Mathur, Rajiv, Doctoral Thesis, Carnegie-Mellon University (1981).
16. Langer, J. S. and H. Müller-Krumbhaar, Phys. Rev. A27 (1983) 499.
17. Dee, G. and J. S. Langer, Phys. Rev. Letter 50 (1983) 383.
18. Langer, J. S., R. F. Sekerka and T. Fujioka, J. Crystal Growth 44 (1978) 414.
19. Langer, J. S. and H. Müller-Krumbhaar, J. Crystal Growth 42 (1977) 11.
20. Langer, J. S. and H. Müller-Krumbhaar, Acta Met. 26 (1978) 1681, 1689, 1697.
21. Mullins, W. W. and R. F. Sekerka, J. Appl. Phys. 35 (1964) 444.
22. Kerszberg, Michel, Phys. Rev. B27 (1983) 3909, 6796.
23. Delves, R. T., J. Crystal Growth 3,4 (1968) 562.
24. Delves, R. T., J. Crystal Growth 8 (1971) 13.
25. Coriell, S. R., M. R. Cordes, W. J. Boettinger and R. F. Sekerka, J. Crystal Growth 49 (1980) 13.
26. Coriell, S. R. and R. F. Sekerka, J. Physico-Chemical Hydrodynamics 2, (1981) 281.
27. Coriell, S. R. and R. F. Sekerka, J. Crystal Growth 34 (1976) 157.
28. Coriell, S. R. and R. F. Sekerka, J. Crystal Growth 61 (1983) 499.

The author was partially supported by National Science Foundation Grant DMR 78-22462.

Mellon College of Science
Carnegie-Mellon University
Pittsburgh, Pennsylvania 15213

DYNAMICS OF FIRST ORDER PHASE TRANSITIONS

M. Slemrod

0. INTRODUCTION

This paper considers continuum thermodynamics of first order phase transitions. Specifically we study the role of viscosity, capillarity, and heat conduction and the relation of these effects to interphase wave propagation.

The paper is divided into six sections. Section 1 imbeds the standard inviscid, adiabatic hydrodynamic equations in a more general van der Waals-Korteweg gradient theory [1],[2] as suggested in the paper of Felderhof [3]. In section 2 we study the zero parameter limit as the coefficients of viscosity, capillarity, and heat conductivity tend to zero. We show that for an "ideal" gas, i.e. if $\frac{\partial p}{\partial w} < 0$ for p the pressure, w specific volume, which is sufficiently viscous, that the limit of solutions of the higher order theory will actually be a solution of the usual inviscid, adiabatic equations. The crucial role of the $\frac{\partial p}{\partial w} < 0$ inequality motivates the conjecture that for a van der Waals type material (for which $\frac{\partial p}{\partial w} > 0$ for some values of specific volume and temperature) this result is not true. In

ISBN 0-12-309770-3

Section 3 we consider isothermal motions and give conditions for the existence of propagating phase boundaries possessing "structure". In Section 4 we show that in the isothermal case the Lax-Friedrichs finite difference scheme induces effects similar to that given by viscosity and capillarity. In Section 5 we show that the full system of equations given in Section 1 possesses traveling wave solutions connecting different phases. Section 6 provides remarks on research in shock splitting and chaos in material exhibiting phase transitions.

The results of Section 3 and 5 have appeared in [4], [5], [6]. The results of Sections 2 and 4 are new.

1. One dimensional Lagrangian description of compressible fluid flow.

We follow the presentation of Courant and Friedrichs [7] of a Lagrangian description of compressible fluid flow based on the law of conservation of mass. The fluid flow is thought of a taking place in a take of unit cross section along the x-axis. Let $\phi(a,t)$ denote position of a fluid particle at time t which had position $x=a$ at time $t=0$. For each X define $x(X,t)$ implicitly by the relationship

$$X = \int_{\phi(0,t)}^{x(X,t)} \rho(s,t)ds \tag{1.1}$$

where $\rho(x,t)$ denotes the density of the fluid at position x and time t. Thus X is the mass of the fluid in the segment of the tube $[\phi(0,t), x(X,t)]$. Differentiation of (1.1) implies $1 = x_X(X,t)\,\rho(x(X,t),t)$ Set $\rho(x(X,x),t) = \tilde{\rho}(X,t)$, $w(X,t) = \tilde{\rho}(X,t)^{-1}$ (the specific volume), $u(X,t) = x_t(X,t)$ (the velocity).

Also we let

p	the pressure
τ	the stress
$\varepsilon > 0$	specific internal energy,
$E \equiv \frac{u^2}{2} + \varepsilon$	
$\tilde{E} \equiv \frac{u^2}{2} + \varepsilon + \frac{A}{2} w_X^2$	(A > 0 constant), specific total energy
h	heat flux
q	specific heat absorption,
b	specific body force,
$\theta > 0$	absolute temperature,
η	specific entropy,
$\Psi \equiv \varepsilon - \theta\eta$	specific Helmholz free energy,
$\mu > 0$	viscosity,
C_V	specific heat at constant volume,
$K > 0$	coefficient of thermal conductivity.

The equations of balance of linear momentum and mass are

$$\rho\ddot{x} = \tau_x + \rho b$$
$$\dot{\rho} + \rho\dot{x}_x = 0 \qquad (1.2)$$

where $\cdot = \frac{d}{dt}$. We now apply the chain rule and rewrite (1.2) in terms of the independent variables X,t to obtain

$$x_{tt} = \tau_X + b,$$
$$(\rho\, x_X)_t = 0, \qquad (1.3)$$

where we have used the fact that $\dot{\rho}(x,t) = \tilde{\rho}_t(X,t)$. (1.3:b) is automatically satisifed since $\rho x_X = 1$.

To write down an equation for balance of energy we follow an idea of B.U. Felderhof [3] (though more recent work in this direction is to be found in [8]). Let us assume τ can be broken up into two contributions: τ_1 due to pressure and viscous contributions, τ_2 due to capillarity effects in spirit of van der Waals [1] and Korteweg [2],[9], i.e.

$$\begin{aligned} \tau_1 &= -p + \mu u_X , \\ \tau_2 &= - A\, w_{XX} , \\ \tau &= \tau_1 + \tau_2 . \end{aligned} \tag{1.4}$$

Felderhof's postulate is that the internal energy is influenced only by the internal stress τ_1 according to the balance law

$$\rho\dot{\varepsilon} = \rho q + \tau_1 \dot{x}_x + h_x \tag{1.5}$$

Again used of the chain rule allows us to rewrite (1.5) as

$$\varepsilon_t = \tau_1\, x_{tX} + h_X + q. \tag{1.6}$$

We make the constitutive hypothesis that the pressure and internal energy are functions of w and θ, $p = p(w,\theta)$, $\varepsilon = \varepsilon(w,\theta)$, so that in the X,t coordinates the balance of momentum, mass, and energy become

$$\boxed{\begin{aligned} u_t &= -p(w,\theta)_X + \mu u_{XX} - A w_{XXX} && \text{(momentum)}, \\ w_t &= u_X && \text{(mass)}, \\ \varepsilon(w,\theta)_t &= (-\,p(w,\theta) + \mu u_X) u_X + h_X . && \text{(energy)}. \end{aligned}} \tag{1.7}$$

We have set $q = b = 0$ for simplicity. Also for simplicity we take μ to be a constant > 0.

If we make use of the field relationships

$$\frac{\partial\psi}{\partial w} = -\,p, \qquad \frac{\partial\psi}{\partial\theta} = -\,\eta \tag{1.8}$$

and the identity

$$\psi_t = \psi_w w_t + \psi_\theta \theta_t = (\varepsilon - \theta\eta)_t$$

we find from (1.7,b) that

$-pu_X = \varepsilon_t = \theta\eta_t$. Hence (1.7c) determines the entropy production law:

$$\theta\eta_t = \mu u_X^2 + h_X . \qquad \text{(entropy prod. law)} \qquad (1.9)$$

From the definition $\tilde{E}$ (the specific total energy) we find

$$\begin{aligned}\tilde{E}_t &= uu_t + \varepsilon_t + Aw_X w_{Xt} \\ &= u(-p_X + \mu u_{XX} - Aw_{XXX}) - pu_X + \mu u_X^2 \\ &\quad + Aw_X u_{XX} + h_X \\ &= -(pu)_X + \mu(uu_X)_X - (Auw_{XX})_X \\ &\quad + Au_X w_{XX} + Aw_X u_{XX} + h_X\end{aligned}$$

i.e.

$$\tilde{E}_t = (u[-p + \mu u_X - Aw_{XX}])_X + (Au_X w_X)_X + h_X .$$
(balance of total energy) (1.10)

From now on we shall assume h is given by Fourier's law

$$h = K\,\theta_X \qquad \text{(Fourier's Law)} \qquad (1.11)$$

$K \geqslant 0$ a constant.

2. The zero parameter limit for the initial value problem

We consider the initial value problem for the balance of momentum (1.7a), mass (1.7b), and total energy (1.10) where h is given by (1.11). For convenience we record these equations here:

$$u_t = -p(w,\theta)_X + \mu u_{XX} - Aw_{XXX} \qquad (1.7a)$$

$$w_t = u_X \qquad (1.7b)$$

$$\tilde{E}_t = (u[-p + \mu u_X - Aw_{XX}])_X + (Au_X w_X)_X + K\theta_{XX} \qquad (1.10)$$

Define $\underset{\sim}{\lambda} = (\mu, A, K)$ and set $|\underset{\sim}{\lambda}| = \lambda$

We prescribe initial data

$$\begin{aligned} u(X,0;\underset{\sim}{\lambda}) &= u_o(X;\underset{\sim}{\lambda}), \\ w(X,0;\underset{\sim}{\lambda}) &= w_o(X;\underset{\sim}{\lambda}), \qquad -\infty < X < \infty, \\ \theta(X,0;\underset{\sim}{\lambda}) &= \theta_o(X;\underset{\sim}{\lambda}). \end{aligned} \qquad (2.1)$$

We assume u_o, w_o, θ_o are smooth, go to zero sufficiently fast as $|X| \to \infty$ and are such that

$$-\int_{-\infty}^{\infty} \eta(w_o(X;\underset{\sim}{\lambda}),\ \theta_o(X,\underset{\sim}{\lambda}))dX \leq \text{const (independent of } \underset{\sim}{\lambda})$$

and $\|u_o(\cdot,\underset{\sim}{\lambda})\|_{L^2(-\infty,\infty)}$, $\|\varepsilon(w_o(\cdot,\underset{\sim}{\lambda}),\theta_o(\cdot,\underset{\sim}{\lambda})\|_{L^1(-\infty,\infty)}$

$A\ \| w_{o_X}(\cdot,\underset{\sim}{\lambda})\|^2_{L^2(-\infty,\infty)}$ are uniformly bounded by some constant independent of $\underset{\sim}{\lambda}$.

We shall be interested in the singular limit of solutions as $\lambda \to 0+$, In particular let us assume for $\mu>0, K>0, A>0$ smooth solutions of (1.7a, 1.7b, 1.10), (2.1), exist on the strip $\Sigma = (-\infty,\infty) \times [0,T]$ for which

$$\begin{aligned} &\| u(\cdot,\cdot,\underset{\sim}{\lambda}) \|_{L\infty(\Sigma)} + \| w(\cdot,\cdot,\underset{\sim}{\lambda}) \|_{L\infty(\Sigma)} \qquad (2.2) \\ &+ \| \theta(\cdot,\cdot,\underset{\sim}{\lambda}) \|_{L\infty(\Sigma)} \leq \text{Const. (independent of } \underset{\sim}{\lambda}) \end{aligned}$$

$$\int_{-\infty}^{\infty} \eta\ (w(x,t;\underset{\sim}{\lambda}),\theta(x,t;\underset{\sim}{\lambda})dX \leq \text{Const. (independent of } \underset{\sim}{\lambda})$$

$$\text{and} \quad u(X,t;\underset{\sim}{\lambda}) \to \bar{u}(X,t),$$
$$w(X,t;\underset{\sim}{\lambda}) \to \bar{w}(X,t), \tag{2.3}$$
$$\theta(X,t;\underset{\sim}{\lambda}) \to \bar{\theta}(X,t), \quad \text{as } \lambda \to 0+,$$

a.e. on Σ.

Question. In what sense (if any) do $\bar{u},\bar{w},\bar{\theta}$ satisfy (1.7a, 1.7b, 1.10) with $\lambda=0$ (i.e. $A = \mu = K = 0$)?

We shall resolve this question through a sequence of integral estimates. In what follows const. means some positive constant independent of λ.

Lemma 2.1 $\mu \|u_X\|^2_{L^2(\Sigma)}$, $K\| \frac{\theta_X}{\theta} \|^2_{L^2(\Sigma)}$,

$$K\| \theta_X\|^2_{L^2(\Sigma)}, \quad A\|w_X\|^2_{L^2(\Sigma)} \le \text{Const.} \tag{2.4}$$

Proof. From (1.9) we see

$$-\eta_t = -\frac{\mu}{\theta}u^2_X - K\left(\frac{\theta_X}{\theta}\right)_X - K\frac{\theta_X^2}{\theta^2} \tag{2.5}$$

so

$$\int_{-\infty}^{\infty} \eta(w,\theta)\Big|_{t=0}^{t=T} dX = \iint_\Sigma \frac{\mu}{\theta} u^2_X + K\frac{\theta_X{}^2}{\theta^2}.$$

Hence

$$-\int_{-\infty}^{\infty} \eta(w(X,T;\underset{\sim}{\lambda}), \theta(w(X,T;\underset{\sim}{\lambda}))dX + \iint_\Sigma \frac{\mu}{\theta} u^2_X + K\frac{\theta_X{}^2}{\theta^2}$$

$$= -\int_{-\infty}^{\infty} \eta(w(X,0;\underset{\sim}{\lambda})\theta(w(x,0;\underset{\sim}{\lambda}))dX$$

which by (2.2) and our assumptions on the initial data imply

$$\iint_\Sigma \frac{\mu}{\theta} u^2_X + \frac{\theta_X^2}{\theta^2} \le \text{Const.}$$

Since θ is positive and bounded from above we see

$$\mu\|u_X\|_{L^2(\Sigma)}, \quad K\|\frac{\theta_X}{\theta}\|_{L^2(\Sigma)}, \quad K\|\theta_X\|^2 \le \text{Const.}$$

Finally from (1.10) we see

$$A\int_{-\infty}^{\infty} w^2_X(X,t;\underset{\sim}{\lambda})dX \le \int_{-\infty}^{\infty} \tilde{E}(X,t)dX = \int_{-\infty}^{\infty} \tilde{E}(X,0)dX \le \text{Const.}$$

and the lemma is proven. ☒

Theorem 2.1 $\bar{u}, \bar{w}, \bar{\theta}$ satisfy

$$\bar{u}_t = -p(\bar{w},\bar{\theta})_X,$$
$$\bar{w}_t = \bar{u}_X, \qquad (2.6)$$
$$\eta(\bar{w},\bar{\theta})_t \geq 0,$$

in the sense of distributions. Furthermore if $A^{1-2\delta} \leq \text{Const.}\ \mu$ for some $0 < \delta < 1$ then

$$\tilde{E}(\bar{u},\bar{w},\bar{\theta})_t + (p(\bar{w},\bar{\theta})\bar{u})_X + f = 0 \qquad (2.7)$$

In the sense of distributions for some unique distribution f. In particular (2.7) will hold when $A = A_o\mu^2$, A_o a positive constant.

Proof. Multiply (2.5) by a non-negative test function $\phi \in C_o^\infty(\Sigma)$. We see

$$-\iint_\Sigma \eta_t\phi \leq \sqrt{K}\iint_\Sigma \phi_X \sqrt{K}\left(\frac{\theta_X}{\theta}\right)$$
$$\leq \sqrt{K}\left(\iint_\Sigma \phi_X^2\right)^{1/2}\left(\iint_\Sigma K\left(\frac{\theta_X}{\theta}\right)^2 dX\right)^{1/2}$$
$$\leq \sqrt{K}\ \text{Const.} \to 0 \text{ as } \lambda\to 0+$$

where we have used (2.4). This proves (2.6c); (2.6a,b) follow from multiplication of (1.7a,b) by test functions, integration by parts, and letting $\lambda \to 0+$.

Next multiply (1.10) by C^∞ test function ϕ.

We note

$$\mu\iint_\Sigma \phi\, uu_X = -\mu\iint_\Sigma \phi_X\left(\frac{u^2}{2}\right) \to 0 \text{ as } \lambda \to 0+,$$

$$K\iint_\Sigma \phi\theta_{XX} = K\iint_\Sigma \phi_{XX}\,\theta \to 0 \text{ as } \lambda \to 0+,$$

$$A\iint_\Sigma \phi_X(uw_{XX}) - \phi_X(u_X w_x) =$$

$$-A\iint_\Sigma \phi_{XX}(uw_X) + 2\phi_X(u_X w_X)$$

$$\leq \sqrt{A}\left(\iint_\Sigma (\phi_{XX}u)^2\right)^{1/2}\left(\iint_\Sigma Aw_X^2\right)^{1/2}$$

$$+ 2A^{\delta} \left(\iint_{\Sigma} A^{1-2\delta} u_X^2 \phi_X^2 \right)^{1/2} \left(\iint_{\Sigma} A w_X^2 \right)^{1/2}$$

$$\leq \sqrt{A} \text{ const. } T + 2A^{\delta} \text{ const. } T \left(\iint_{\Sigma} \mu u_X^2 \right)^{1/2}$$

(if $A^{1-2\delta} \leq$ Const. μ).

By (2.4) the last quantity in the above inequality is bounded by Const. $(\sqrt{A} + A^{\delta})$ which goes to zero as $\lambda \to 0+$. Hence we find

$$-\iint_{\Sigma} \phi_t \tilde{E}(\bar{u},\bar{w},\bar{\theta}) - \iint_{\Sigma} \phi_X (p(\bar{w},\bar{\theta})\bar{u}) =$$

$$\lim_{\lambda \to 0+} - \iint_{\Sigma} \phi \left(\frac{A}{2} w_X^2 (X,t;\underset{\sim}{\lambda}) \right)_t . \qquad (2.8)$$

Now $\iint_{\Sigma} \phi \left(\frac{A}{2} w_X^2(X,t;\underset{\sim}{\lambda}) \right)_t$

defines a sequence of distributions which by (2.8) possesses a finite and definite limit for each C^{∞} test function ϕ. Hence from the Banach-Steinhaus theorem (see e.g. [10, p.54, Thm. 2.1]) we know there exists a unique distribution f satisfying

$$\lim_{\lambda \to 0+} \iint_{\Sigma} \phi \left(\frac{A w_X^2}{2} (X,t;\underset{\sim}{\lambda}) \right)_t = \iint_{\Sigma} \phi f \qquad (2.9)$$

This concludes the proof of the theorem. ☒

From (2.9) we see f represents the rate of change of interfacial energy for the "inviscid" $\lambda=0$ system. If $f \equiv 0$ (2.6),(2.7) reduces to the classical equations of compressible inviscid fluid flow. We will now show that in the usual "hyperbolic" case $(p_w < 0)$, $f \equiv 0$ if A,μ are suitably restricted.

<u>Theorem 2.2</u> Assume $p_w(w,\theta) < 0$ and that A,μ,K satisfy the <u>strong dissipation</u> assumption: $K \geq c_1 \mu$, $c_1 > 0$; $c_2 A \geq \mu^2 \geq 4A$ for $c_2 \geq 4$. Then $\bar{u},\bar{w},\bar{\theta}$ satisfy (2.6),(2.7) with $f \equiv 0$.

<u>Proof</u> Define $v(X,t,\underset{\sim}{\lambda}) = u(X,t;\underset{\sim}{\lambda}) - D_2\, w_X(X,t;\underset{\sim}{\lambda})$

where $\begin{matrix} D_1 \\ D_2 \end{matrix} = \frac{\mu}{2} \mp 1/2\, (\mu^2 - 4A)^{1/2}$.

The strong dissipation assumption implies $D_1, D_2 > 0$. Notice $D_1 D_2 = A$, $\mu - D_1 = D_2$, and $A^{1-2\delta} \leq$ const. μ with $\delta = 1/4$, const. $= 1/2$. If we substitute for u in (1.7a,b) we find

$$v_t = -p(w,\theta)_X + D_1\, v_{XX}, \tag{2.10}$$

$$w_t = v_X + D_2\, w_{XX}. \tag{2.11}$$

Now we observe

$$\int_{-\infty}^{\infty} \left(\varepsilon(w,\theta) + \frac{v^2}{2}\right)_t dX =$$

$$\int_{-\infty}^{\infty} - p\, u_X + \mu u_X^2 - p_X v - D_1 v_X^2\, dX =$$

$$\int_{-\infty}^{\infty} -p(v_X + D_2 w_{XX}) + \mu u_X^2 - p_X v - D_1 v_X^2\, dX =$$

$$\int_{-\infty}^{\infty} D_2\, p_w w_X^2 + D_2\, p_\theta w_X\, \theta_X + \mu u_X^2 - D_1 v_X^2\, dX.$$

So for γ a positive constant we find using (2.5) that

$$\int_{-\infty}^{\infty} \left(\varepsilon(w,\theta) + \frac{v^2}{2} - \gamma\eta\right)_t dX =$$

$$\int_{-\infty}^{\infty} \Big(D_2\, p_w\, w_X^2 + D_2\, p_\theta w_X \theta_X + \mu u_X^2 - D_1 v_X^2$$

$$- \frac{\mu\gamma}{\theta}\, u_X^2 - K\gamma\, \frac{\theta_X^2}{\theta^2}\Big)\, dX.$$

Since θ is bounded above, say by $\theta_M > 0$, a constant, we see for

$\frac{\gamma}{\theta_\mu} \geq 1$ that $\int_{-\infty}^{\infty} \left(\varepsilon\,(w,\theta) + \frac{v^2}{2} - \gamma\eta\right)_t dX \leq$

$$\int_{-\infty}^{\infty} \Big(D_2\, p_w\, w_X^2 + D_2\, p_\theta\, w_X\, \theta_X - K\gamma\, \frac{\theta_X^2}{\theta_M^2}\Big)\, dX. \tag{2.12}$$

Since (w,θ) lies in a bounded set of $R \times R^+$ a.e. $p_w(w,\theta) < -a < 0$ for some positive constant a a.e. and $|p_\theta(w,\theta)| \leqslant b$ for some positive constant b a.e. Thus

$$\text{r.h.s. } (2.12) \leqslant \int_{-\infty}^{\infty} - D_2 a w_X^2 + D_2 b |w_X \theta_X| - \frac{K\gamma}{\theta_M^2} \theta_X^2 dX$$

$$\leqslant \int_{-\infty}^{\infty} - D_2 a w_X^2 + D_2 \left(\frac{\nu^2 w_X^2}{2} + \frac{b^2 \theta_X^2}{2\nu^2} \right) - \frac{K\gamma}{\theta_M^2} \theta_X^2 dX$$

$$\leqslant \int_{-\infty}^{\infty} - D_2 w_X^2 \left(a - \frac{\nu^2}{2}\right) - \theta_X^2 \left(\frac{K\gamma}{\theta_M^2} - \frac{D_2 b^2}{2\nu^2}\right) dX$$

for any $\nu \neq 0$. In particular if $\nu^2 = a$ we find

$$\text{r.h.s. } (2.12) = \int_{-\infty}^{\infty} - \frac{a}{2} D_2 w_X^2 - \theta_X^2 \left(\frac{K\gamma}{\theta_M^2} - \frac{D_2 b^2}{2a}\right) dX$$

Now since $D_2 \leqslant \mu$, $k \geqslant c_1 \mu$ we see

$$\text{r.h.s. } (2.12) \leqslant \int_{-\infty}^{\infty} - \frac{a}{2} D_2 w_X^2 - \theta_X^2 \mu \left(\frac{c_1 \gamma}{\theta_M^2} - \frac{b^2}{2a} \right) dX$$

Hence for $\gamma > \max \{\theta_M, \frac{b^2 \theta_M^2}{2ac_1}\}$ we have

$$\text{r.h.s. } (2.12) \leqslant - \frac{a}{2} \int_{-\infty}^{\infty} D_2 w_X^2 dX.$$

Hence from (2.12) we have

$$\int_{-\infty}^{\infty} \varepsilon(w(X,T;\underset{\sim}{\lambda}), \theta(X,t,\underset{\sim}{\lambda})) + \frac{v^2}{2} (X,t;\underset{\sim}{\lambda})$$

$$+ \gamma\eta(w(X,T,\lambda),\theta(X,T;\lambda)) dX$$

$$+ \frac{a}{2} \iint_\Sigma D_2 w_X^2 \leqslant \int_{-\infty}^{\infty} [\varepsilon(w_o(X;\lambda), \theta_o(X,\lambda))$$

$$+ (u_o (X;\underset{\sim}{\lambda}) - D_2 w_{o_X}(X;\underset{\sim}{\lambda}))^2 - \gamma\eta(w_o(X\cdot\lambda), \theta_o(X,\lambda))] dX$$

So by our assumptions on the initial data and by (2.2) we have

$$\frac{a}{2} \iint_\Sigma D_2 w_X^2 \leqslant \text{Const.} + 2D_2^2 \int_{-\infty}^{\infty} w_{o_X} (X;\underset{\sim}{\lambda})^2 dX.$$

Now since $D_2 = \frac{\mu}{2} + 1/2\ (\mu^2 - 4A)^{1/2}$,

$$D_2^2 \leqslant \mu^2 - 2A \leqslant (C_2 - 2)\ A \leqslant 2A.$$ So by our assumption on the initial data that

$$A\int_{-\infty}^{\infty} w_{o_X}(X;\underset{\sim}{\lambda})^2\ dX \leqslant \text{const.}$$

we find $$\iint_\Sigma D_2\ w_X^2 \leqslant \text{const.} \qquad (2.13)$$

Now let us return to the r.h.s. of (2.8)

$$-\iint_\Sigma \phi\ (\frac{A}{2}\ w_X^2(X,t;\underset{\sim}{\lambda}))_t =$$

$$\frac{A}{2}\iint_\Sigma \phi_t\ w_x^2\ (X,t;\underset{\sim}{\lambda}) = \frac{A}{2}\ D_2^{-1}\iint_\Sigma \phi_t D_2\ w_X^2(X,t,\underset{\sim}{\lambda}).$$

Again since $D_2 = \frac{\mu}{2} + 1/2\ (\mu^2 - \psi A)^{1/2} \geqslant \frac{\mu}{2}$,

$D_2^2 \geqslant \frac{\mu^2}{4}$ so $D_2^2 \geqslant A$ (by the strong dissipation assumption). Thus we find

$$|\iint_\Sigma \phi\ (\frac{A}{2}\ w_X^2(X,t;\underset{\sim}{\lambda}))_t| \leqslant$$

$$\sqrt{A}\ \text{Const.} \iint_\Sigma D_2 w_X^2\ (X,t;\underset{\sim}{\lambda}) \to 0$$

as $\lambda \to 0+$ by (2.13). Hence $f \equiv 0$ and the theorem is proven. $\boxed{X}$

Corollary 2.1 Assume $p_w\ (w,\theta) < 0$ and that A,μ,K satisfy $K = K_o\mu$, K_o positive constant; $A = A_o\mu^2$, A_o positive constant, $A_o \leqslant 1/4$, Then $(\bar{u},\bar{w},\bar{\theta})$ satisfy (2.6),(2.7) with $f \equiv 0$.

Proof In this case the strong dissipation assumption holds.

Remark : Notice the strong dissipation assumption excludes the purely dispersive case $A>0$, $\mu=K=0$, and the purely diffusive $A=0$, $\mu>0$, $K>0$.

The importance of the above results is the following: For an "ideal" like fluid i.e. one for which $p_w(w,\theta) < 0$ and if the strong dissipation assumption is satisfied the singular limit (u,w,θ) (if it exists) as $\lambda \to 0+$ will satisfy the classical inviscid equations of compressible fluid flow (2.6, 2.7 with f=0). On the other hand if p_w is not always negative (e.g. van der Waals fluid) we cannot be sure (2.7 with f=0) is satisfied. Hence while (1.7a, 1.7b, 1.10) may be reasonable model for compressible fluid $(A>0, \mu>0, K>0)$ without sign restrictions on p_w, the singular case $A = \mu = K = 0$ may in fact be a poor model unless we restrict $p_w < 0$. Of course in the isothermal case, where we assume the energy equation (1.10) is satisfied identically through addition of a heat source q, we are only interested in the equations of balance of mass and momentum. Theorem 2.1 shows these are satisfied by u,w without sign restrictions on p_w. Hence in the isothermal case we expect (2.6a,b) to provide a reasonable model for compressible fluid flow without sign restrictions on p_w. In this case (2.6a,b) would have to be supplemented by "admissibility criteria" which could be motivated by the "viscous" $A>0, \mu>0$ equations (1.7a,b). This has been done in [4,5] and will be discussed in the next section. Finally we note the recent results of Lax and Levermore [11] leads us to conjecture that in the purely dispersive case $(\mu=0, K=0)$ weak limits of solutions as $A \to 0+$ will exist and will <u>not</u> satisfy the zero parameter equations, even in the purely hyperbolic case $p_w > 0$, for sufficiently long time intervals.

3. Isothermal admissibility criteria

In the previous section we have seen that the isothermal balance laws of mass and momentum

$$u_t = -p(w,\theta)_X \quad , \ \theta > 0 \text{ constant}, \qquad (2.6a,b)$$
$$w_t = u_X$$

can be expected to produce a model of compressible fluid flow consistent with our zero parameter limit theory.

Of course, as is well known, for the case $p_w < 0$, $p_{ww} \neq 0$, (2.6a,b) is a geniunely nonlinear (in the sense of Lax) system of hyperbolic conservation laws and will develop shocks for arbitratily smooth choices of initial data. Furthermore in the case $p_w > 0$ in some region (α, β), (2.6a,b) will be an elliptic initial value problem (at least initially) for initial data $w = w_o \ \varepsilon \ (\alpha, \beta)$. In this case (2.6a,b) loses well-posedness due to the classical Hadamard instability.

We shall be concerned here with the van der Waals fluid whose isotherms are given by

$$p(w,\theta) = \frac{R\theta}{w-b} - \frac{a}{w^2}, \quad 0 < b < w < \infty, \ R, \ a > 0 \qquad (3.1)$$

Typical isotherms of (3.1) are shown in Fig. 3.1.

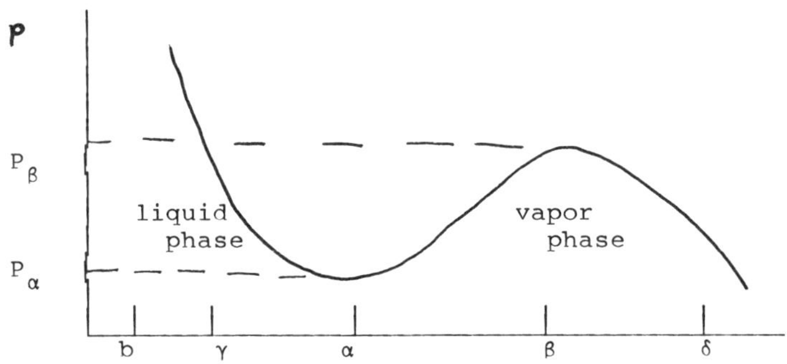

Fig. 1

From Fig. 1 we see that if $\theta < \theta crit = 8a/27br$ for a certain range of pressures $p, p_\alpha < p < p_\beta$ the van der Waals fluid can simultaneously exist in liquid and vapor phases. Furthermore we see $p(\gamma) = p(\beta)$, $p(\alpha) = p(\delta)$,

$p_w < 0$ for $w \in (b,\alpha) \bigcup (\beta,\infty)$ (2.6a,b is hyperbolic),

$p_w > 0$ for $w \in (\alpha,\beta)$ (2.6a,b is elliptic),

and in either region we can expect loss of smooth solutions. One way to deal with this difficulty is to expand our horizons and seek weak solutions of (2.6a,b). Unfortunately weak solutions to the initial value proble for quasi-linear hyperbolic equations are notoriously non-unique. This motivates us to use our discussion in Section 2 to seek admissibility criteria which will hopefully pick out the physically meaningful solutions.

One natural approach is to consider (1.7a,b) as being the physically correct model and admit only those solutions of (2.6a,b) which are limits of solutions of (1.7a,b)

Definition 3.1 A weak solution $(\bar{u},\bar{w})$ of (2.6a,b) is admissible w.r.t (1.7) if it is the distribution limit as $A \to 0+$, $\mu \to 0+$ for $(X,t) \in RxR^+$ of solutions of the problem (1.7a,b; $A>0, \mu>0$).

As an example consider the weak solution of (2.6a,b) given by the equilibrium solution

$u=0, \bar{w}=\bar{w}_\ell$, $X < X_o$, $\bar{w} = \bar{w}_v$, $X > X_o$,

where $\bar{p} = p(\bar{w}_\ell,\theta) = p(\bar{w}_v,\theta)$. (Multiplication of (1.7a,b) by C^∞ test functions and integration by parts yields the weak form of (1.7a,b). A straight forward computation shows the choice of $\bar{u},\bar{w}$ is indeed a weak solution of (2.6a,b).

Now let us examine the equilibrium solutions of (1.7a,b) given by $u=0$, $w=w(X)$, where

$0=-p(w,\theta)_X - Aw_{XXX}$, $-\infty < X < \infty$

If we require $w \to \bar{w}_\ell$ as $X \to -\infty$, $w \to \bar{w}_v$ as $X \to +\infty$, we find $0 = -p(w,\theta) + \bar{p} - Aw_{XX}$, $-\infty < X < \infty$.

Now set $\xi = \left(\frac{X - X_o}{A^{1/2}}\right)$, $w(X) = \hat{w}(\xi)$.

Then we see

$$0 = -p(\hat{w},\theta) + \bar{p} - \hat{w}''(\xi) \tag{3.2}$$

where $' = \frac{d}{d\xi}$. To solve (3.2) let $w'(\xi)=y(w(\xi))$.

Then

$$0 = -p(\hat{w},\theta) + \bar{p} - 1/2\frac{d}{d\hat{w}}(y^2)$$

so

$$y^2(\hat{w}) = -2\int_{w_\ell}^{\hat{w}} (p(s,\theta) - \bar{p})ds. \tag{3.3}$$

Hence $\hat{w}(\xi)$ is obtained as the solution of $\frac{d\hat{w}(\xi)}{d\xi} = y(\hat{w}(\xi))$,

$\hat{w}(-\infty) = w_\ell$, where y is given by (3.3). For $\hat{w}(+\infty) = \bar{w}_v$

We need $y(w_v) = o$ i.e.

$$\int_{w_\ell}^{w_v}(p(s,\theta) - \bar{p})\, ds = 0; \tag{3.4}$$

this is the Maxwell equal area rule (see Fig.2)

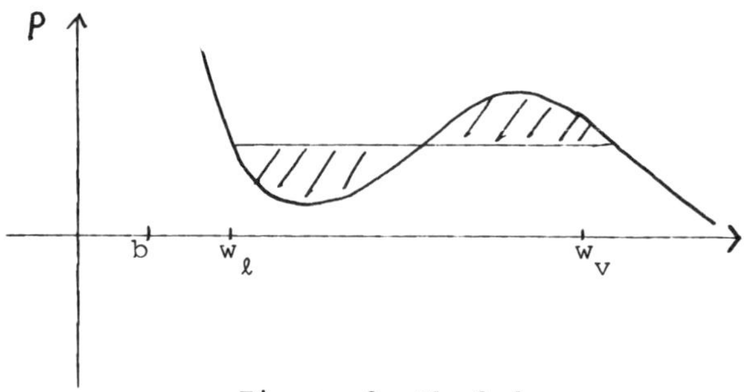

Figure 2 Shaded areas are equal.

Hence we see if $\bar{w}_\ell, \bar{w}_v$ satisfy Maxwell's rule, $w(X) \to \bar{w}_\ell$, $X < X_o$, $w(X) \to \bar{w}_v$, $X > X_o$, and $\bar{w}$ is admissible w.r.t. (1.7)

Another approach to admissibility is to use (2.10), (2.11) as our underlying "viscous" system. This motivates the following definition.

__Defn. 3.2__ A weak solution $(\bar{u},\bar{w})$ of (2.6a,b) is __admissible w.r.t (2.10), (2.11)__ if it is the distributional limit as $D_1, D_2 \to 0+$ for $X,t \in R \times R+$ of solutions (v,w) of the problem (2.10, 2.11;
$D_1 > 0, D_2 > 0$).

__Theorem 3.1__

(i) If $(\bar{u},\bar{w})$ is admissible w.r.t (1.7) and $\mu^2 > 4A$ then (u,w) is the distributional limit of solutions (v,w) of (2.10), (2.11) where

$$D_1 = \frac{\mu}{2} \mp \tfrac{1}{2}\sqrt{\mu^2 - 4A} \quad \text{or} \quad D_1 = \frac{\mu}{2} \pm \tfrac{1}{2}\sqrt{\mu^2 - 4A}$$
$$D_2 \qquad\qquad\qquad\qquad\qquad D_2$$

(ii) If $(\bar{u},\bar{w})$ is admissible w.r.t (2.10),(2.11) then $(\bar{u},\bar{w})$ is the distributional limit of solutions (u,w) of (1.7a,b) where $A = D_1 D_2$, $\mu = D_1 + D_2$.

__Proof__ (i) Consider (1.7) and set
$v(X,t) = u(X,t) - D_2 w_X(X,t)$. Then (v,w) satisfies (2.10),(2.11) for the specified D_1, D_2.
If $u \to \bar{u}$, $w \to \bar{w}$ then $w_X \to \bar{w}_X$, $D_2 w_X \to 0$ in and hence $v \to u$ (all convergence in the sense of distributions).

(ii) Consider (2.10),(2.11) and set $u(X,t) = v(X,t) + D_2 w_X(X,t)$. Then (u,w) satisfies (1.7) with $\mu = D_1 + D_2$, $A = D_1 D_2$. The result now follows in a manner analagous to (i).

As there is no straight forward manner to test whether a given weak solution of (2.6a,b) satisfies either of these criteria, we present a variant of these criteria based on the special case of the Riemann inital value problem. Consider the initial value problem for (2.6a,b) with initial data

$$u = u- \quad, X < 0; \quad u = u+ \quad, X > 0; \qquad (3.5)$$
$$w = w- \qquad\qquad w = w+$$

where $u-, w-, u+, w+$ satisfy the Rankine-Hugoniot jump conditions

$$-U[u] = -[p] \qquad (3.6)$$
$$-U[w] = [u]$$

for some U, $[u] = u+ - u-$, etc. Then

$$u = u- \quad, X < Ut; \qquad u = u+ \quad, X > Ut; \qquad (3.7)$$
$$w = w- \qquad\qquad w = w+$$

is a weak solution of (2.6a,b,3.5). The propagating singular surface $X = Ut$ is a shockwave. Furthermore for the van der Waals fluid if $w_- \varepsilon\ (b,\alpha)$, $w_+ \varepsilon\ (\beta,\infty)$ or vice versa the shockwave is a <u>propagating</u> <u>phase</u> <u>boundary</u>.

Motivated by our discussion of admissibility of weak solutions we attempt to approximate (3.7) by solutions of (1.7) and (2.10) (2.11). In particular for (1.7) set $A = A_0\mu^2$ and look for traveling wave solutions of the form

$$u = u(\zeta), \quad w = w(\zeta), \quad \zeta = \frac{X-Ut}{\mu} \quad .$$

We find w satisfies

$$A_o\hat{w}'' + U\hat{w}' + p(\hat{w},\theta) - p(w-) + U^2(\hat{w}-w-) = 0, \qquad (3.8)$$
$$\hat{w}(-\infty) = w- \;, \qquad \hat{w}(+\infty) = w_+ . \qquad (3.9)$$

Defn. 3.3 If (3.8), (3.9) possesses a solution we say (3.7) satisfies the viscosity - capillarity shock criterion.

Similarly if we look for traveling wave solutions

$$v = \hat{v}(\zeta), \quad w = \hat{w}(\zeta), \quad \zeta = \frac{X - Ut}{D_1 + D_2}$$

of (2.10),(2.11) we find

$$\frac{D_1 D_2}{(D_1+D_2)^2} \hat{w}'' + U\hat{w}' + p(\hat{w},\theta) - p(w-) + U^2 (\hat{w}-w-) = 0, \tag{3.10}$$

$$\hat{w}(-\infty) = w-, \quad \hat{w}(+\infty) = w+.$$

Defn. 3.4 If (3.10) possesses a solution we say (3.7) satisfies the viscosity shock criterion with viscosity matrix diag (D_1, D_2).

Trivially we observe the following theorem.

Theorem 3.2 The shock criteria of Defns. 3.3 and 3.4 are equivalent for $A_o = \frac{D_1 D_2}{(D_1 + D_2)^2}$;

Now for the van der Waals fluid and w_-, U given where w_- (b,α) let w_+ (U) denote the solution (if it exists) of the Rankine-Hugoniot condition

$$U^2 = \frac{-[p]}{[w]} \tag{3.11}$$

lying in the interval (β,∞). Also set

$$f_-(\xi;U) = U^2(\xi-w-) + p(\xi,\theta) - p(w-),$$

$$f_+(\xi \cdot U) = U^2(\xi-w_+) + p(\xi,\theta) - p(w_+).$$

Lemma 3.1 Let w_- be given in (b,α) and assume either

$b < w- < \gamma$

or

$$\text{(I)} \quad \int_{w-}^{w_+(0)} f-(\xi;o)d\xi < 0.$$

Then there exists a unique value U^*, $0 < U^* < U$, such that (3.8) possesses a solution with $w(-\infty) = w-$, $w(+\infty) = w_+(U^*)$, if U such that

$$\int_{w-}^{w+(\bar{U})} f_-(\xi,\bar{U})d\xi = 0.$$

Similarly if for $w_+ \varepsilon(\beta,\infty)$ given we define $w-(U)$ to be the solution (if it exists) of (3.11) in the interval (b,α).

Lemma 3.2 Let w_+ be given in (β,∞) assume either $w_+ > \delta$ or

(II) $\int_{w-(0)}^{w_+} f_+(\xi;o)\, d\xi > 0.$

Then there exists a unique value U^*, $0 < U^* < U$, such that (3.8) possesses a solution with $\hat{w}(-\infty) = \hat{w}-(U^*)$, $w(+\infty) = w_+$, if U such that

$$\int_{w-(\bar{U})}^{w_+} f_+(\xi;\bar{U})d\xi = 0$$

Remark. The hypotheses of Lemma 3.1 have a simple interpretation. Either $b < w_- < \gamma$ in which case $w_+(0)$ doesn't exist, or $\gamma < w_- < \alpha$ and (I) holds. (I) says that the signed area between the chord joining $(w-,p(w-,\theta))$ and $(w_+(0), p(w_+(o),\theta))$ and the graph of $p(w,\theta)$ between w_- and $w_+(0)$ is negative. U is that positive value of U such that the signed area between the chord joining $(w_-,p(w-,\theta)$ and $(w_+(\bar{U}), p(w_+(\bar{U}),\theta)$ and the graph of $p(w,\theta)$ between $w-$ and $w_+(\bar{U})$ is zero. (See Fig 3.)

Analagous interpretations hold for Lemma 3.2.

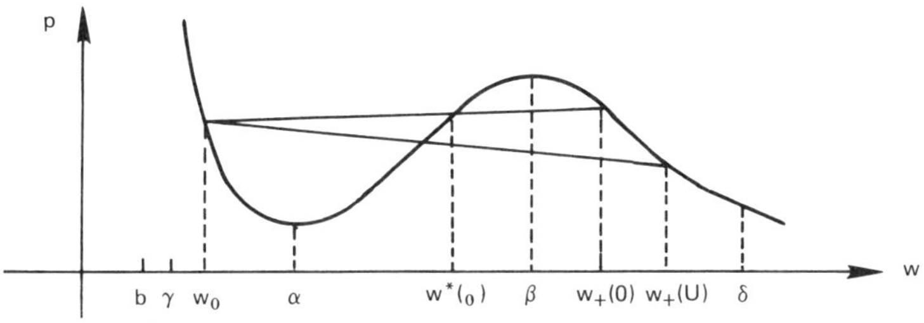

Figure 3

Proof of Lemma 3.1 Let $y(w(\zeta)) = w'(\zeta)$ where we have dropped the ^ symbol. Then (3.8) may be rewritten as

$$\frac{Ao}{2} \frac{d}{dw} y^2(w) + Uy(w) + f_-(w;U) = 0.$$

Integration from w- to w yields

$$\frac{Ao}{2} y^2(w) = - U \int_{w-}^{w} y(\xi)d\xi - \int_{w-}^{w} f(\xi\cdot U)d\xi. \quad (3.12)$$

Let us consider the case $U = 0$. Assume for the moment $\gamma \leq w- < \beta$ so that $w_+(0)$ exists. Let $w^*(0)$ denote that value of w for which the chord connecting $(w-,p(w-,\theta))$ and $(w_+,p(w_+(0),\theta))$ intersects the graph of $p((w,\theta))$, $\alpha < w^*(0) < \beta$ See Figure 3. Since $f_-(\xi,0) > 0$ for $w^*(0) < \xi < w_+(0)$ it follows that

$$\frac{Ao}{2} y^2(w) \geq - \int_{w-}^{w_+(0)} f_-(\xi;0)d\xi > 0$$

by (I) . Hence if $\gamma \leq w- < \beta$ and (I) holds the solution of (3.9) for $U = 0$ with $w(-\infty) = w-$ always stays above the line $y=0$. On the other hand if $b < w- < \gamma$, $f_-(\xi;0) < 0$ for $\xi > b$ and by (3.12) the solution of (3.9) for $U = 0$, $w(-\infty) = w-$, always stays above $y=0$.

Now set $U_\ell = \sup\{U';\ y(w^+(U))$ exists and is positive for $0 \leq U < U'\}$. For $U = \bar{U}$ we have by (3.12) that

$$\frac{Ao}{2} \quad y^2(w) = -\bar{U}\int_{-w}^{w_+(U)} y(\xi)d\xi.$$

Thus $y(w)$ must go from positive to negative values as w varies from w- to $w_+(\bar{U})$ See Figure 4. Hence $0 < U_\ell < \bar{U}$.

Now let $0 < U_n < U_\ell$, $U_n \to U_\ell$ as $n \to \infty$. The sequence $\{y(w_+(U_n))\}$ is bounded and hence has a convergent subsequence $y(w_+(U_{n_k})) \to y_o$ as $k \to \infty$. Thus $y(w_+(U_\ell)) = y_o$ by continuity with respect to the parameter U. If $y_o > 0$ then by continuity with respect to U $y(w_+(U)$ exists for $U_\ell + \varepsilon > U > U_\ell$ and $\varepsilon > 0$ sufficiently small since orbits have $w'(\xi) > 0$ in the upper half w-y plane, $w > b$, (the trajectories can't turn around). If $y_o < 0$ then by continuity with respect to U $y\ (w_+\ (U)) < 0$ for some $0 < U < U_\ell$ contradicting the definition of U_ℓ.
Hence $y_o = 0$ and $y(w_+(U_\ell)) = 0$. U_ℓ is the $\bar{U}^*$ of the lemma. ☒

The proof of Lemma 3.2 is similar.

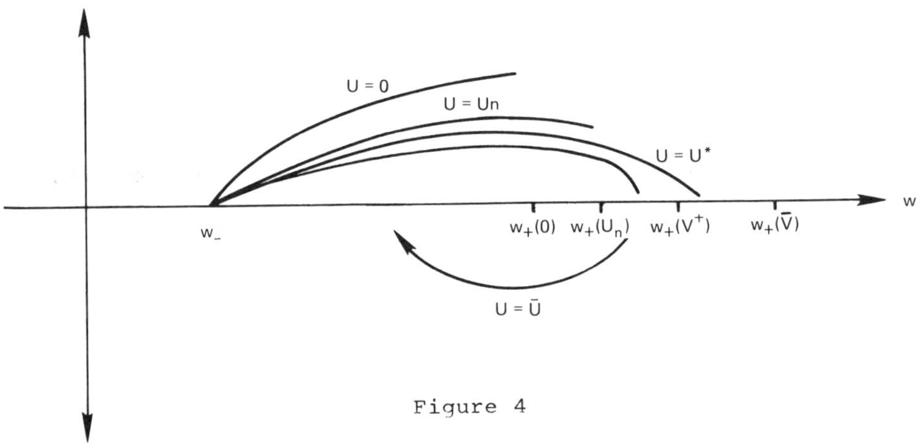

Figure 4

From the above lemmas and the Rankine-Hugoniot jump conditions we immediately obtain the following theorem.

Theorem 3.3 (i) Give a constant state (u-,w-), $w_- \varepsilon(b,\alpha)$, the propagating phase boundary (3.7) with

$$w_+ = w_+(U^*) \ , \ u_+ = u_- -U^* \ (w_+(U^*) - w-)$$

satisfies the viscosity - capillarity shock admissibility criterion if the hypotheses of Lemma 3.1 are satisfied.

(ii) Given a constant (u_+,w_+), $w_+ \ \varepsilon \ (\beta,\infty)$, the propagating phase boundary (3.7) with

$w_- = w_-(U^*)$, $u_- = u_+ + (w_+ - w_- \ (U^*))$ satisfies the viscosity - capillarity shock admissibility criterion if the hypotheses of Lemma 3.2 are satisfied.

We note that Theorem 3.3 says that a homogeneous state in one phase determines both a second homogenous state in the second phase and a speed of propagation so as to make (3.7) admissible with respect to the viscosity-capillarity condition. This stands in contract to classical fluid mechanics ($p_w < 0$) where both the state one side of a shock and the speed of propagation determines the speed on the other side of the shock.

Finally we note for the static shock case $U = 0$ the computation of the transition layer was made by van der Waals in his classic paper [1]. Since then the equilibrium problem has been considered by many others. Good references are Rowlinson's introduction to [1;1973], and Widom's paper [12]. Recent views on the equilibrium equal area rule have been presented by Serrin [13] and Aifantis and Serrin [14].

4. Finite differences

In Section 3 we have seen that in the isothermal case introduction of viscous and capillarity terms in (2.6a,b) provides a way of picking out admissible weak solutions consistent with the mechanical principle outlined in Sections 1 and 2. Motivated in part by the paper of Lax [15] we shall see in this section that discretization of (2.6a,b) has an effect similar to that of introduction of viscosity and capillarity.

In particular consider the Lax-Friedrichs scheme for (2.6a,b) given by

$$\frac{1}{\Delta t}\left\{ u(X,t+\Delta t). - \frac{u(X+\Delta X,t)+u(X-\Delta Xt;)}{2}\right\} +$$

$$\frac{1}{2\Delta x}\{p(w(X+\Delta X,t)) - p(w(X-\Delta X,t))\} = 0,$$

$$\frac{1}{\Delta t}\left\{w(X,t+\Delta t) - \frac{w(X+\Delta X,t)+w(X-\Delta X,t)}{2}\right\}$$

$$-\frac{1}{2\Delta x}\{u(X+\Delta X,t) - u(X-\Delta X,t)\} = 0,$$

where we have suppressed the parametric dependence of p on θ.

Set $\lambda = \frac{\Delta t}{\Delta X}$. If (4.1) has smooth solutions u,w we may expand u,w in Taylor series about X,t. From (4.1) we find

$$u_t + p(w)_X = \frac{\Delta t}{2}(\phi(w,\lambda)u_X)_X + 0(|\Delta t|^2),$$

$$w_t - u_X = \frac{\Delta t}{2}(\phi(w,\lambda)w_X)_X + 0(|\Delta t|^2), \qquad (4.2)$$

where $\phi(w,\lambda) = p_w(w) + \lambda^{-2}$.

From (4.2) we see (4.1) is first order accurate with respect to the equation (2.6a,b) but second order accurate with respect to

$$\begin{aligned} u_t + p(w)_X &= \frac{\Delta t}{2} \quad (\phi(w,\lambda)u_X)_X , \\ w_t - u_X &= \frac{\Delta t}{2} (\phi(w,\lambda)w_X)_X , \end{aligned} \tag{4.3}$$

which a non-constant viscosity matrix version of (2.15), (2.11).

In order to force well-posedness on (4.3) a natural assumption is $\phi(w,\lambda) > 0$ which makes (4.3) parabolic. In regions where $p_w > 0$ (for example $w \varepsilon (\alpha,\beta)$ in the van der Waals fluid) $\phi(w,\lambda) > 0$ automatically. In hyperbolic regions where $p_w < 0$ we need.

$$\lambda^2 \max \{ - p_w ; w \ni p_w < 0\} < 1 \tag{CFL}$$

which is just the Courant-Friedrichs-Lewy condition. CFL will hold if λ is sufficiently small. Hence if CFL is satisfied (4.3) is a well posed parabolic system and should the scheme (4.1) converge for some fixed λ (4.1) will provide a <u>second order</u> accurate approximation of (4.3)

As we have observed in Section 3 viscosity matrix criteria and viscosity - capillarity criteria can be connected through a change of dependent variables. So it comes as no suprise that the parabolic system (4.3) will have properites reminiscent of (1.7a,b).

For example if we wish to see how (4.3) approximates a shock front (3.7) for (2.6a,b) we could look for traveling wave solutions for (4.3) of the form

$$u(X,t) = \hat{u}(\xi), \ w(X,t) = \hat{w}(\xi), \quad \xi = \frac{X - Ut}{\Delta t} .$$

We easily find $\hat{w}$ satisfies

$$\phi(\phi w')' + U\phi w' + U^2(w-w_-) - p(w) + p_- = 0 \quad (4.4)$$

$$w(-\infty) = w_-, \; w(+\infty) = w_+, \quad (4.5)$$

where $p_- = p(w_-,\theta)$ and we have dropped the ^ .

Set $\phi w' = y(w(\xi))$. Then $\phi(\phi w')' = y'(w(\xi)y(w(\xi))$ and (4.4) becomes $\frac{1}{4}\frac{d}{dw}\ (y^2(w)) + Uy + U^2(w-w_-)\ - p(w) + p_- = 0$ (4.6)

which is just (3.8) with $A_0 = \frac{1}{4}$. Hence we expect that for Δt small and CFL satisfied the difference scheme (4.1), if it converges, will produce shocks consistent with viscosity-capillarity condition.

In particular for the state shock $U = 0$ an analysis identical to that for (3.2) shows that we expect (4.1) to reproduce only static shocks that approximate the Maxwell equal area rule.

Notice that unlike the results of Section 3 this is as a consequence of our difference method (4.1) and not because of any law of physics or constitutive assumptions.

Needless to say these remarks motivate the possibility of playing the above scenario in reverse: If we wish to approximate solutions of (2.6a,b) satisfying the viscosity-capillarity condition the method (4.1) is a natural choice. We note λ may have to be reduced further if we desire to satisfy inequality (1.26) of Lax [15] regarding his "entropy" inequality ((1.13) of [15]).

5. Traveling wave solutions connecting liquid and vapor phases in a van der Waals fluid

In this section we return to our equations of balance of momentum, mass, and energy (1,7a,b), (1.10) which include the effects of viscosity, capillarity, and heat conduction. Also we assume we are dealing with a van der Waals fluid which satisfies constitutive equation (3.1). We recall from Fig. 1 that for a typical non-monotone isotherm

(i) $p_w(w,\theta) < 0$, $0 < b < w < \alpha$, $\beta < w$,

(ii) $p_w(\alpha,\theta) = p_w(\beta,\theta) = 0$,

(iii) $p_w(w,\theta) > 0$ if $\alpha < w < \beta$.

The domains (b,a) and (p,∞) are called the α (or liquid phase) and β (or vapor phase) respectively.

Assume w_- is a constant value of the specific volume lying in the α-phase of a $\theta_- < \theta_{crit}$ isotherm. Our goal will be to see how a homogeneous equilibrium state of (1.7a,b), (1.10) $w = w_-$, $u = u_-$, $\theta = \theta_-$ associated with the fluid being liquid can be dynamically transferred to (or from) a second equilibrium state $w = w_+$, $u = u_+$, $\theta = \theta_+$ associated with the fluid being vapor. Mathematically this means we shall try to find a traveling wave solution of (1.7a,b), (1.10) connecting equilibrium states (w_-,u_-,θ_-) and (w_+,u_+,θ_+) where w_- is in the α-phase of the θ_- isotherm and w_+ is in the β-phase of the θ_+ isotherm.

In order to simplify our calculations we take

$$A = A_o\mu^2 \text{ , } A_o \text{ a positive constant ,}$$

in (1.7a,b), (1.10).

We seek a traveling wave solution of the form

$$u = \hat{u}(\xi) \ , \ w = \hat{w}(\xi) \ , \ \theta = \hat{\theta}(\xi) \ , \ \xi = \frac{X - Ut}{\mu} \qquad (5.1)$$

(U the speed of the traveling wave) subject to

$$\hat{u}(-\infty) = u_- \ , \ \hat{w}(-\infty) = w_- \ , \ \hat{\theta}(-\infty) = \theta_- . \qquad (5.2)$$

Substitution of (5.1) into (1.7a,b), (1.10) yields the system

$$\begin{aligned} -Uu' &= (-p + u' - A_o w")' \ , \\ -Uw' &= u' \ , \qquad (5.3) \\ -U\tilde{E}' &= [u(-p + u' - A_o w")]' + A_o(u'w')' + \frac{K}{\mu}\theta" \ , \end{aligned}$$

where $' = \frac{d}{d\xi}$ and the ^ 's have been dropped.

Set $\varepsilon_- = \varepsilon(w_-,\theta_-)$, $p_- = p(w_-,\theta_-)$, $\tilde{E}_- = \frac{u_-^2}{2} + \varepsilon_-$, and integrate (5.3) from $-\infty$ to ξ . We then find u,w,θ satisfy the system

$$\begin{aligned} -U(u-u_-) &= -(p-p_-) + u' - Aw" \ , \\ -U(w-w_-) &= u-u_- \ , \qquad (5.4) \\ -U(\tilde{E}-\tilde{E}_-) &= u(-p + u' - A_o w") + u-p_- + A_o u'w' + \frac{K}{\mu}\theta' \ . \end{aligned}$$

Finally we use (5.3.b) and (5.4b) to eliminate u and u' from (5.4a,c). A straightforward computation shows (w,θ) satisfy the equations

$$\begin{aligned} w' &= v \ , \\ A_o v' &= -Uv - U^2(w - w_-) - p(w,\theta) + p_- \ , \qquad (5.5;U,\mu) \\ \theta' &= \frac{\mu}{K} U\{-(\varepsilon(w,\theta)-\varepsilon_-) - p_-(w - w_-) + \frac{U^2}{2}(w - w_-) + A_o v^2\} \end{aligned}$$

In terms of $(5.5;U,\mu)$ our goal outlined at the beginning of this section becomes the following: Find a solution of $(2.6;U,\mu)$ connecting the equilibrium points

$w = w_-$, $v = 0$, $\theta = \theta_-$, $w = w_+$, $v = 0$, $\theta = \theta_+$ of $(5.5;U,\mu)$ where w_- is in the α-phase of the θ_- isotherm and w_+ is the β-phase of the θ_+ isotherm.

The reader familiar with the Fitzhugh-Nagumo equations for nerve impulse transmission may notice that $(5.5;U,\mu)$ is remarkable similar to the system

$$\begin{aligned} W' &= V , \\ A_o V' &= -UV + G(W, \Theta) , \\ \Theta' &= \mu U^{-1} H(W, \Theta) , \end{aligned} \qquad (FN - TW)$$

governing traveling wave solutions of the Fitzhugh-Nagumo equations. One difference between (FN - TW) and $(5.5;U,\mu)$ is that in $(5.5;U,\mu)$ the equilibrium points depend on U , in (FN -TW) they don't. Nevertheless the two sets of equations are sufficiently alike to that a modification of Carpenter's argument [16] proving the existence of a heteroclinic orbit for (FN - TW) will yield existence of a solution to $(5.5;U,\mu)$ connecting α and β phases.

First let us recall some thermodynamics. For the van der Waals fluid the Helmholtz free energy satisfying (1.8) is given by

$$\boxed{\psi(w,\theta) = -R\theta \,\ell n(w - b) - \frac{a}{w} + F(\theta)}$$

where $F(\theta)$ is an arbitrary function θ . Since $\varepsilon = \psi + \theta\eta$ it follows that

$$\varepsilon(w,\theta) = -\frac{a}{w} + F(\theta) - \theta F'(\theta) .$$

As a simplifying assumption we set (See [17], p. 74) $F(\theta) = -C_V \theta \ell n \theta$ + constant where c_V is a positive constant to obtain

$$\boxed{\varepsilon(w,\theta) = -\frac{a}{w} + c_V \theta + \text{constant.}}$$

The constant c_V is the specific heat at constant volume.

Now in addition to our earlier constitutive assumptions we shall assume

(i) μ the viscosity is small,

(ii) c_V is large.

While on physical grounds K should be small as well as μ our goal here is to isolate the effects of comparitively smaller contributions of viscosity and capillarity as opposed to thermal conductivity. In other words we are considering μ/K the Prandtl number to be small.

Since μ is assumed to be small, a useful first step in the analysis of $(5.5;U,\mu)$ is to study $(5.5;U,o)$. This has been done earlier in Lemmas 3.1 and 3.2. For example Lemma 3.1 assures the existence of an isothermal solution of $(5.5;U^*(\theta_-),0)$ connecting $(w_-,0,\theta_-)$ to $(w_+(U^*(\theta_-)),0,\theta_-)$. Unfortunately this solution does not satisfy the full system $(5.5;U^*(\theta_-),\mu)$ when $\mu \neq 0$. The reason is, of course, that the quantity in braces in $(5.5;\ U^*(\theta_-),\mu{:}c)$ would have to be identically zero along this isothermal solution and it is not.

We also know that if μ is small and U is near $U^*(\theta_-)$ solutions of $(5.5;U,\mu)$ stay close to solutions of $(5.5,U^*(\theta_-),0)$ provided w' or v' is not small. If w'

and v' are small and $\mu > 0$, then the "slow" system (5.5;U,μ;c) becomes "fast" relative to the "fast" system (5.5;U,μ:a,b). A <u>singular solution</u> of (2.6;U,μ) will consist of alternating solutions or solution segments of the two systems (5.5), $U^*(\theta_-),0)$ and (5.5,$U^*(\theta_-),\mu$:c), the latter being defined where $w' = v' = 0$, i.e. on $w = g(\theta)$, a solution of

$$-U^*(\theta_-)^2(w - w_-) - p(w,\theta) + p(w_-,\theta_-) = 0 . \qquad (5.6)$$

Our hope is that if we can construct a singular solution of (5.5;U,μ) connecting the state $w = w_-$, $v = 0$, $\theta = \theta_-$ in the (liquid) α-phase to a state $w = w_+$, $v = 0$, $\theta = \theta_+$ in the (vapor) β-phase we can find a true solution fo (5.5;U,μ) that does the same thing, provided μ is sufficiently small.

From Lemma 3.1 we know the first piece of our singular solution will be a curve in the $\theta = \theta_-$, $v > 0$ half plane connecting $(w_-,0,\theta_-)$ to $(w_+(U^*(\theta_-)),0,\theta_-)$. According to our plan the next piece of the singular solution will be to follow the flow of (5.5;$U^*(\theta_-),\mu$;c) when $w = g(\theta)$. Examination of the van der Waals isotherms (Figure 1) for differing values of θ leads to the graph of g shown in Figure 5.

To follow the solution of (5.5,$U^*(\theta_-),\mu$,c) along $w = g(\theta)$ means studying the scalar equation

$$\theta' = \mu \frac{U^*(\theta_-)}{K} \{-(\varepsilon(g(\theta),\theta) - \varepsilon_-) - p_-(g(\theta)-w_-) + \frac{U^*(\theta_-)^2}{2}(g(\theta) - w_-)^2\} . \qquad (5.7)$$

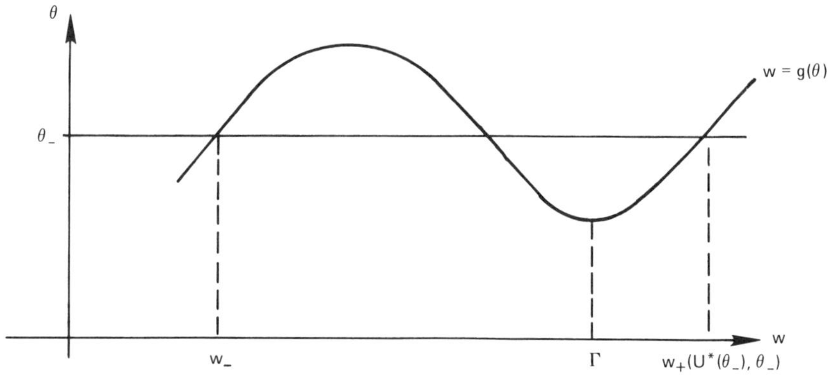

Figure 5

As the first part of the singular solution took us from $(w_-,0,\theta_-)$ to $(w_+(U^*(\theta_-)),0,\theta_-)$, it is necessary to study the behavior of (5.7) only on the $w > \Gamma$ branch of $g(\theta)$ where $g'(\theta) > 0$ (Figure 5).

To study the dynamics of (5.7) we must first identify the equilibrium points. These are found as the intersections of the graph $w = g(\theta)$ and $w = \ell(\theta)$ where $w = \ell(\theta)$ is the solution of

$$-(\varepsilon(w,\theta) - \varepsilon_-) - p_-(w - w_-) + \frac{U^*(\theta_-)^2}{2}(w - w_-)^2 = 0 . \tag{5.8}$$

If we substitute our constitutive relation $\varepsilon(w,\theta) = -\frac{a}{w} + c_V\theta + \text{constant}$ into (5.8) and use the fact that

$$U^*(\theta_-)^2 = \frac{-p(w_+(U^*(\theta_-),\theta_-) + p_-}{w_+(U^*(\theta_-)) - w_-} \tag{5.9}$$

we find $w = \ell(\theta)$ satisfies

$$c_V(\theta - \theta_-) = a\left(\frac{1}{w} - \frac{1}{w_-}\right) \qquad (5.10)$$

$$- \left(\frac{w - w_-}{w_+(U^*(\theta_-)) - w_-}\right)\{p_-(w_+(U^*(\theta_-)) - \frac{w}{2} - \frac{w_-}{2})$$

$$+ p(w_+(U^*(\theta_-),\theta_-)\left(\frac{w - w_-}{2}\right)\} .$$

Differentiation of (5.10) with respect to w yields

$$c_V \frac{d\theta}{dw} = \frac{-a}{w^2} - \left(\frac{1}{w_+(U^*(\theta_-)) - w_-}\right)\{[w_+(U^*(\theta_-)) - w]p_- \qquad (5.11)$$

$$+ (w - w_-)p(w_+(U^*(\theta_-)),\theta_-)\}.$$

We see if the θ_- isotherm is as shown in Figure 1 with $p(w_+(U^*(\theta_-)),\theta_-) > 0$, $p_- > 0$, we have $\frac{d\theta}{dw} < 0$ for $w_- < w < w_+(U^*(\theta_-))$ and $w = \ell(\theta)$ is a decreasing function of θ on the range $w_- < w < w_+(U^*(\theta_-))$. From (5.11) we then see if c_V is sufficiently large the curves $w = g(\theta)$ and $w = \ell(\theta)$ will intersect at three points in $w - \theta$ plane as shown in Figure 6 giving the equilibrium points of (5.7). Thus (5.7) has an equilibrium point $\theta = \tilde{\theta}_+$ so that $w_+ = g(\tilde{\theta}_+) > \Gamma$ and $g'(\tilde{\theta}_+) > 0$.

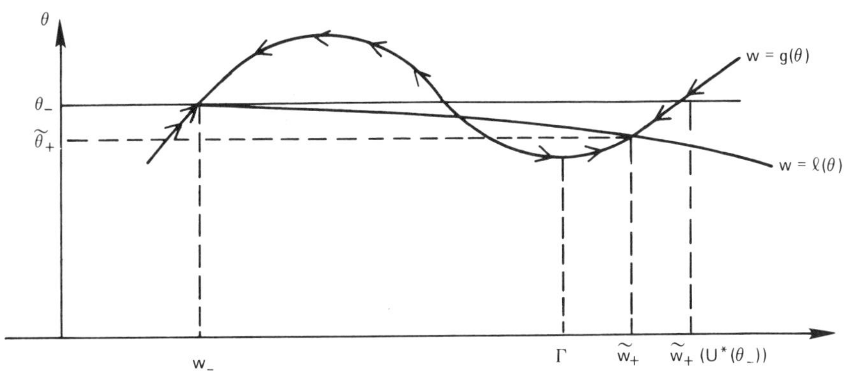

As noted in the preceding paragraph we need $p(w_+(U^*(\theta_-),\theta_-) > 0$. Examination of Lemma 3.1 shows this will be true <u>if</u>: $p(w_+(\bar{U}(\theta_-),\theta_-) > 0$ <u>where</u> $\bar{U}(\theta_-)$ <u>is as defined in Lemma 3.1.</u> We also note the result depends on c_V being sufficiently large. Otherwise the curve $\ell(\theta)$ might fail to intersect $g(\theta)$ except at the point (w_-,θ_-). For this reason we make the constitutive assumption: c_V <u>is large enough to (i) force</u> ℓ g <u>to intersect in three points as shown in Figure 6 and (ii) yield</u> $p(w_+,\theta_+) > 0$.

<u>Lemma 5.1</u>. The equilibrium point $\tilde{\theta}_+$ (where $\tilde{w}_+ = g(\tilde{\theta}_+) > \Gamma$) of (5.7) is asymptotically stable.

<u>Proof</u>. From Taylor's theorem we know

$$\theta' = \mu \frac{U^*(\theta_-)}{K} \{-\varepsilon_w(\tilde{w}_+,\tilde{\theta}_+) - p(\tilde{w}_+,\tilde{\theta}_+)\} g'(\tilde{\theta}_+)(\theta - \tilde{\theta}_+)$$

$$- \mu \frac{U^*(\theta_-)}{K} \varepsilon_\theta(\tilde{w}_+,\tilde{\theta}_+)(\theta - \tilde{\theta}_+) + O(|\theta - \theta_+|^2) . \tag{5.12}$$

where we have used the Rankine-Hugoniot relation (3.11). We note that $\varepsilon_\theta = c_V > 0$, $\varepsilon_w = \frac{a}{w^2} > 0$. Secondly our assumption on c_V preceding the statement of the lemma implied $p(\tilde{w}_+,\tilde{\theta}_+) > 0$. So $\tilde{\theta}_+$ is asymptotically stable. The full phase flow of (5.1) is shown in Figure 6 by the arrows on the graph of g.

If we combine Lemma 3.1 and 5.1 we see we have constructed a singular solution of $(5.5;U,\mu)$. The singular solution consists of

(i) an isothermal solution of $(5.5;U^*(\theta_-),0)$ running from $(w_-,0,\theta_-)$ to $(w^*(U^*(\theta_-)),0,\theta_-)$ in the $v > 0$, $\theta = \theta_-$ half plane;

(ii) a solution of (5.7) traveling on the graph of g in the $v = 0$ plane from the point $(w_+(U^*(\theta_-)),0,\theta_-)$ to the equilbrium of (5.7) $\tilde{\theta}_+$. Pictorially the singular solution is represented in Figure 7.

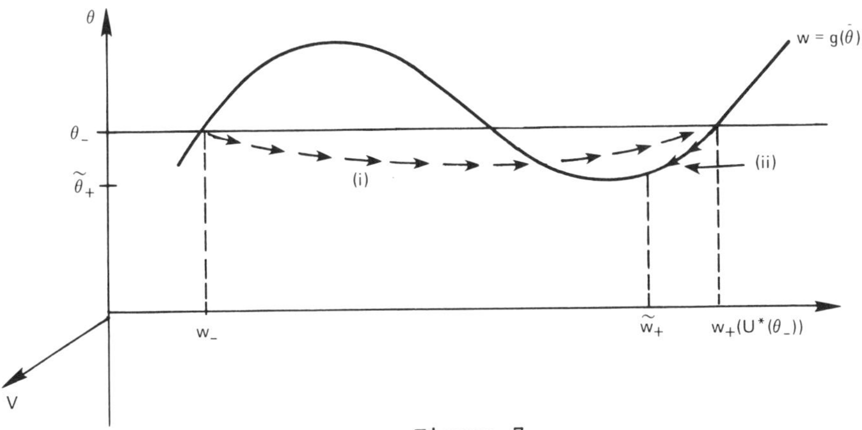

Figure 7

As w_- is in the θ-phase of the θ_- isotherm and $\tilde{w}_+$ is the β-phase of the θ_+ isotherm we see this singular solution does indeed connect the α and β phases.

Having constructed the singular solution our next goal would be to show there is a true solution of $(5.5;U,\mu)$ for U near $U^*(\theta_-)$ and μ small with the same connecting properties as the singular solution. To do this requires use of the Conley-Easton theory of isolating blocks [18]. A reasonable description of the isolating block theory and its application to the problem at hand is beyond the scope of this survey. The interested reader may consult [6] for details. We record here the final result.

Theorem 5.1 (i) (Compression Wave: vapor $\mapsto$ liquid) Assume w_-, θ_-, u_- are given; w_-, $U(\theta_-)$, satisfy the hypotheses of Lemma 3.1. Furthermore assume c_V is sufficiently large and $p(w_+(\overline{U}(\theta_-),\theta_-) > 0$. Then for μ sufficiently small there exists $U_\mu > 0$ and a solution of $(5.5;U_\mu,\mu)$ with $w(-\infty) = w_-$, $v(-\infty) = 0$, $\theta(-\infty) = \theta_-$, $w(+\infty) = w_+^*(U_\mu)$,
$v(+\infty) = 0$, $\theta(+\infty) = \theta_+^*(U_\mu) < \theta_-$; $w_+^*(U_\mu)$ is the β-phase of $\theta^*(U_\mu)$ isotherm. Furthermore, $U_\mu \to U^*(\theta_-) > 0$,

$w^*(U_\mu) \to \tilde{w}_+$, $\theta^*(U_\mu) \to \tilde{\theta}_+ < \theta_-$. as $\mu \to 0+$.

(ii) (Expansion Wave: liquid $\mapsto$ vapor) Assume w_+, θ_+, u_+ are given; w_+, $\overline{U}(\theta_+)$, satisfy the hypotheses of Lemma 3.1. Furthermore assume c_V is sufficiently large and $p(w_+,\theta_+) > 0$. Then for μ sufficiently small there exists $U_\mu < 0$ and a solution of $(5.5;U_\mu,\mu)$ with
$w(-\infty) = w^*(U_\mu)$, $v(-\infty) = 0$, $\theta(-\infty) = \theta^*(U_\mu) > \theta_+$,
$w(+\infty) = w_+$, $v(+\infty) = 0$, $\theta(+\infty) = \theta_+$; $w^*(U_\mu)$ is in the α-phase of the $\theta^*(U_\mu)$ isotherm. Furthermore
$U_\mu \to U^*(\theta_+) < 0$, $w^*(U_\mu) \to w_-$, $\theta^*(U_\mu) \to \theta_- > \theta_+$ as $\mu \to 0_+$
where $\tilde{w}_-$, $\tilde{\theta}_-$ are obtained in a manner analogous to that given for the construction of $\tilde{w}_+$, $\tilde{\theta}_+$.

Two observations are in order. First note that Theorem 5.1 predicts that the wave speeds $U^*(\theta_-)$ in (i) and $U^*(\theta_+)$ in (ii) will provide a good approximation to the true wave speeds. Thus while the isothermal equation $(5.5;U,0)$ yields the wrong equilibrium states to which a transition is made it does yield a good approximation to the correct speed of transition.

Secondly, we note that in the experiment of Dettleff, Thompson, Meier and Speckman [19] a wave was produced which yielded complete liquefaction of super heated vapor i.e. a complete transition from metastable superheated vapor to liquid. The liquids used in their study were of "retrograde type" (their terminology) in that they possessed high specific heat at constant volume (e.g. fluorocarbons). They also noted an increase in temperature from vapor to liquid phases These observations are consistent with Theorem 5.1 (i).

6. Final Remarks

In this survey we have touched on various aspects of one dimensional wave propagation in materials exhibiting phase transitions. Due to limitations of space two other topics of both mathematical and physical interest have been omitted. First we have not discussed more complicated shock dynamics and in particular the issue of shock "splitting". In the paper of Hagan and Slemrod [5] it has been shown that the "structure" theory described in Section 3 may be used to predict shock "splitting" of the vapor $\mapsto$ liquid phase transition. This is consistent both with physical arguments suggested by Zel'dovich and Raizer [20] and the recent experimental results of Thompson and Kim [21]. Secondly we have not described the recent work of Marsden and Slemrod [22] relating the continuum theory (with a small periodic in space and time heat source) and the Mel'nikov-Holmes-Marsden approach to chaos. In particular [22] shows how the continuum model of Section 1 predicts chaotic spinodal decomposition (in time) and chaotic interfaces (in space).

REFERENCES

1. van der Waals, J.D. [1893], Veshandel. Konik. Akad. Weten. Amsterdam. vol. 1, No. 8; [1979], Translation of J.D. van der Waals' "The Thermodynamic theory of capillarity under the hypothesis of a continuous variation of density" by S. Rowlinson, J. Statistical Physics 20, 197-244.
2. Korteweg, D.J. [1901], Sur la forme que prennent les equations du mouvement des fluides si l'on tient compte des forces capillaires par des variations de densite, Archives Neerlandaises des Sciences Exactes et Naturelles.
3. Felderhof, B.U. [1970], Dynamics of the diffuse gas-liquid interface near the critical point, Physica 48, 541-560.
4. Slemrod, M. [1983], Admissibility criteria for propagating phase boundaries in a van der Waals fluid, Archive for Rational Mechanics and Analysis 81, 301-315.
5. Hagan, R. and M. Slemrod [1984], the viscosity-capillarity criterion for shocks and phase transitions, Archive for Rational Mechanics and Analysis 83, 333-361.
6. Slemrod, M. [1984], Dynamic phase transitions in a van der Waals fluid, to appear J. Differential Equations.
7. Courant, R. and K.O. Friedrichs [1948], Supersonic Flow and Shock Waves. New York: John Wiley.

8. Dunn, J.E., and J. Serrin [1983], On the thermodynamics of interstitial working, prepring #24 Institute for Mathematics and its Applications, University of Minnesota, Minneapolis, Minnesota 55455, to appear Archive for Rational Mechanics and Analysis.

9. Truesdell, C.A. and W. Noll [1965], The non-linear field theory of mechanics, Vol. III/3 of the Encyclopedia of Physics, S. Flugge, editor. Heidelberg, New York: Springer.

10. Mizohata, S. [1973], The theory of partial differential equations. London: Cambridge University Press.

11. Lax, P.D. and C.D. Levermore [1983], The small dispersion limit of the Korteweg-deVries equations, Comm. Pure and Applied Mathematics 36, p. 253-290.

12. Widom, B. [1977], Structure and thermodynamics of interfaces, in "Statistical mechanics and statistical methods in theory and application", ed. U. Landman, New York: Plenum.

13. Serrin, J. [1980], Phase transitions and interfacial layers for van der Waals fluids, Proc. SAFA IV Conference, Recent Methods in Nonlinear Analysis and Applications, Naples, March 21-28, 1980. A Canfora, S. Rionero, C. Sbordone, C. Trombetti, editors.

14. Aifantis, E. and J. Serrin, [1982], Towards a mechanical theory of phase transformations, Technical report, Corrosion Research Center, University of Minnesota, Minneapolis, Minnesota 55455.

15. Lax, P.D., [1975], Shock waves and entropy, in Contributions to Nonlinear Functional Analysis, ed. E.H. Zarantonello, p. 603-635, New York: Academic Press.

16. Carpenter, G. [1977], A geometric approach to singular perturbation problems with applications to nerve impulse equations, J. Diff. Equations 23, p. 335-367.

17. Fermi, E. [1956], Thermodynamics, New York: Dover.

18. Conley, C. and R. Easton [197], Isolated invariant sets and isolating blocks, Trans. Amer. Math. Soc. 158, p. 35-61.

19. Dettleff, G., Thompson, P.A., Meier, G.E.A., and H.-D. Speckman [1979], An experimental study of liquefaction shock waves, J. Fluid Mechanics 95, p. 279-304.

20. Zel'dovich, Ya. B. and Yu. P. Raizer [1966], Physics of shock waves and high-temperature hydrodynamic phenomena, New York: Academic Press.

21. Thompson, P.A. and Y.-G. Kim [1983], Direct observation of shock splitting in a vapor-liquid system, to appear Physics of Fluids.

22. Slemrod, M. and J. Marsden [1983], Temporal and spatial chaos in a van der Waals fluid due to periodic thermal perturbations, preprint #33 Institute for Mathematics and its Applications, Univ. of Minnesota, Minneapolis, Minnesota 55455, to appear in Advances in Applied Mathematics.

ACKNOWLEDGEMENT: I would like to thank the institute for Mathematics and its Applications, University of Minnesota, for their kind hospitality during my 1983 visit. Also I would like to thank Prof. R. DiPerna for his valuable remarks.

This research was performed in part while the author was a Senior member, Institute for Mathematics and its Applications, University of Minnesota, Minneapolis, MN 55455, and supported at I.M.A. by grants from AFOSR and NSF. The research was also supported in part by the Air Force Office of Scientific Research, Air Force Systems Command, USAF, under contract/grant no. AFOSR-81-0172. The United States Government is authorized to reproduce and distribute reprints for government purposes notwithstanding any copyright notation hereon.

Department of Mathematical Sciences
Rensselaer Polytechnic Institute
Troy, NY 12181

EQUILIBRIUM SHAPES OF SURFACES AND GRAIN BOUNDARIES

J. E. Taylor

The surface free energy component of the total free energy affects the local shape of the boundaries of single crystals, whether the boundaries be interfaces with other crystals, with a fluid, or with a vapor, in much the same way that surface free energy affects the shapes of soap films and soap bubbles. The main differences are that the surface free energy per unit surface area (hereafter called the **surface tension**) of solids is likely to be anisotropic, and that equilbrium is much less likely to be attained for solids than fluids for kinetic reasons. In this paper, some of the geometric effects are given of minimizing the surface free energy (with a given anisotropic surface tension function). The questions of how this surface tension function was determined and of what role kinetics plays in determining geometries are not treated. Similary, "surface stresses"--that is, all elastic energies--are ignored.

1. SURFACE TENSION FUNCTIONS.

The surface tension is taken to be a given function

$$F: \mathbf{S}^2 \to \mathbf{R}^+;$$

$\mathbf{S}^2$ is the unit sphere and represents the space of possible oriented unit normals to a surface, and $\mathbf{R}^+$ is the positive real numbers. (The surface tension function is often called γ or σ in other works.) Thus the surface (free) energy $\mathbf{F}(S)$ of an oriented polyhedral surface S is simply the sum over each of the polyhedral faces of S of the area of the face times the value of F on its oriented unit normal. More generally, for any oriented surface having an oriented unit normal $\nu_S(x)$ at H^2 almost all x in S,

PHASE TRANSFORMATIONS AND MATERIAL INSTABILITIES IN SOLIDS

ISBN 0-12-309770-3

$$\mathbf{F}(S) = \int_{x \in S} F(\nu_S(x))\, dH^2x$$

(here H^2 is Hausdorff 2-dimensional area, which agrees with any reasonable definition of surface area where such area is well-defined and additionally gives a precise meaning to area in surfaces with complicated singularities).

The information contained in a surface tension function is quite usefully presented by the result W_F of Wulff's construction, which will be called the **crystal** of F and abbreviated by W when F is clear by context. This shape, whose surface has the least surface energy for a solid of the same volume (see [13], [7], and [10], for example), is

$$W_F = \{x \in \mathbf{R}^3 \colon x \cdot z \leqslant F(\nu) \text{ for every } \nu \in \mathbf{S}^2\}.$$

(If we consider the inside of W to be of phase I and the outside of W to be of phase II, then we are implicitly choosing the orientation of the normal of a surface to point from phase I to phase II; the central inversion of W is the equilibrium shape of a body of phase II submerged in a matrix of phase I, and thus even if W has a center of symmetry, the surface of the central inversion of W has the opposite orientation to that of W.)

Note that if F is a constant (the isotropic case), then W_F is a ball, as expected. F is **elliptic** if and only if W_F is uniformly convex (has positive upper and lower bounds on its curvatures); F is called **crystalline** if and only if W_F is a polyhedron (note, however, that crystalline solids need not have crystalline surface tension functions, particularly if the temperature is high). F is **convex** if and only if for any function G with $G \leqslant F$, $W_G = W_F$ implies $G = F$. In figure 1a, a W_F is shown that will be used for figure 2 below. In figure 1b, W_F is shown in cross section, along with the cross section of the polar plot of a nonconvex F which has W_F as its crystal. In figure 1c the polar plot of the convex integrand with this crystal is shown (in cross section). Observe that if F

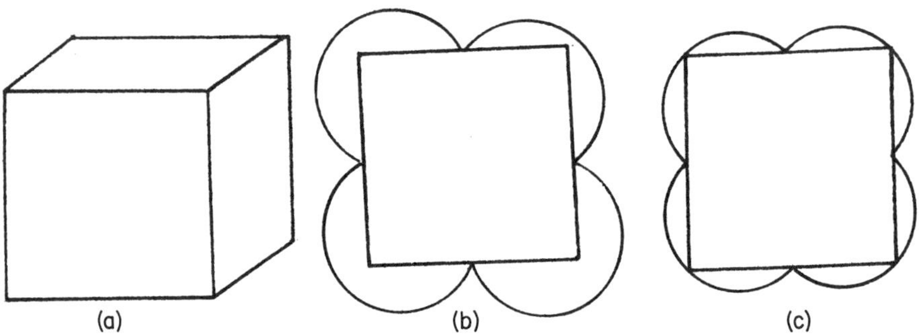

Figure 1. A Wulff shape, and illustrations of cross-sections of nonconvex and convex functions producing it.

is convex and crystalline, then it is determined by its values on a finite number of points, the directions of the faces of W_F.

2. THE PRESCRIBED BOUNDARY PROBLEM.

The major problem addressed here is the **prescribed boundary problem**: to determine the surface(s) of least surface energy spanning a given boundary. Although this precise problem may not be an experimentally reasonable one, one can think of it as isolating a part of a larger interface for the purpose of determining what local structures are energy minimizing. There are implicit problems contained in the prescribed boundary problem: do solutions always exist? are they unique? how smooth are they? what is the structure of singularities (places where the surfaces are not smooth)? and finally, how might one compute the minimizing surface(s), given a (reasonably nice) curve as boundary? In fact, what definition does one use for a surface and for a boundary, and what does it mean for a surface to span a boundary?

Here, we take a surface to be part of the boundary of an open set in $\mathbf{R}^3$, and consider only boundaries that are piecewise C^1 curves. (More general definitions are considered in section 3A below.)

2A. CONSTANT SURFACE TENSION FUNCTIONS.

For the case of isotropic surface tension ($F \equiv$ a constant), the prescribed boundary problem is known as **Plateau's problem**. In the above formulation, the local structure of minimizing surfaces is trivial: solutions are smooth everywhere (in fact they are analytic). The global behavior may be quite complicated, however, in that they may have many handles (for example, phase-antiphase domains are highly interconnected).

2B. CRYSTALLINE SURFACE TENSION FUNCTIONS.

We now turn our attention to crystalline surface tension functions. There is an immediate problem with proving any smoothness results. It can be shown that with a prescribed boundary B consisting of a triangle in a plane parallel to one cutting off one corner off of W, that part of the boundary of W which was cut off is an F-minimizing surface with boundary B [10]. There are, however, uncountably many other surfaces of the same surface energy on that prescribed boundary, including one with a fractal-type structure as shown in figure 2a. On the other hand, if the boundary B is a rectangle in a plane which is parallel to an edge of a crystal and which cuts off two corners of the crystal, and if F is nonconvex, then there can be <u>no</u> classical surfaces of least surface energy, since any surface not in the plane of the boundary can have its energy decreased by a deformation which makes it closer to a surface such as the one illustrated in figure

Figure 2. For W as in figure 1a, (a) an **F**-minimizing surface; (b) a boundary with no **F**-minimizing surface, together with a surface whose energy is close to the infimum.

2b (one which is close to the plane but with corrugations parallel to the long parts of W which were cut off). The part of the plane bounded by this rectangle, on the other hand, may not be of least energy because the value of F on this direction could be quite large. The answer to this problem of nonexistence is either to work only with convex surface tension functions, in which case that planar surface is **F**-minimizing, or to extend the notion of what a "surface" is to include "infinitesimally corrugated" surfaces. One way of making that precise is the notion of **varifolds** (see [1] for a complete definition), which are similar to the generalized surfaces of L. C. Young [14].

The philosophy now is to use the finiteness of the number of faces of W. One would like to be able to show that the set of normal directions to a surface should be only those of W (and perhaps a controlled number of others), and to determine all possible local structures (**F**-minimizing ways that plane segments with these orientations can fit together along edges and at corners). One would then know the local structure of minimizing surfaces (and thereby perhaps almost locally minimizing surfaces). Furthermore, if one could additionally know how the local structures can piece together combinatorially, then the total surface energy for a given combinatorial structure would become merely a quadratic function of the distances of the various planes from a fixed reference point, and the exactly minimizing surfaces could be computed. With some appropriate additional hypotheses, this has in fact been accomplished for certain convex piecewise affine boundaries; see in particular [11].

The first key to the understanding of what local structures are minimizing is to see what kinds of surfaces consisting of two half-circles meeting along a common diameter (and oriented so that their union is an oriented surface with no boundary down that diameter) are minimizing. It turns out [10] that, if one looks only at plane segments whose normals are normals of W, and if W has only three faces meeting at each corner, there are only two general types: surfaces which are translations of subsets of the surface of W (i.e. the two plane segments are parallel to faces which are adjacent in W, and the normals to the faces point away from each other), and surfaces which are translations of subsets of the surface of the central inversion of W (here the normals point towards each other). Intersections of the first type are called **regular**, and intersections of the second type are called **inverse**.

The types of corners made up of plane segments parallel to faces of W which turn out to be **F**-minimizing are illustrated in figure 3; they can all be shown to be minimizing by using the corners of W or its central inversion as "barriers" (see

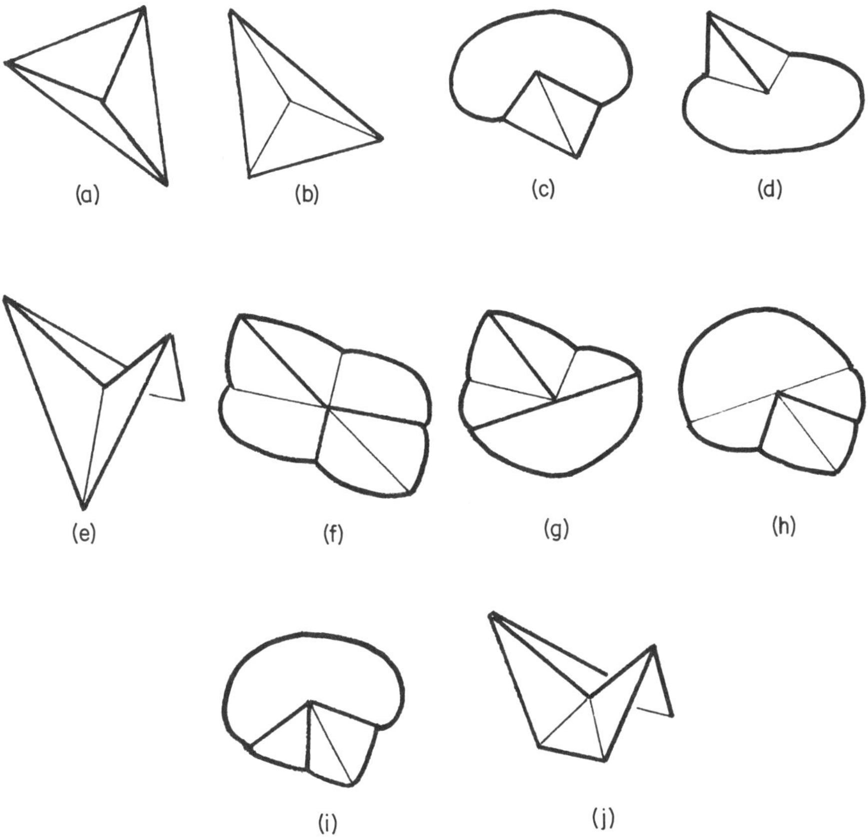

Figure 3.The basic types of **F**-minimizing corners. Regular edges are shown dark and inverse shown lighter.

[5]). The first four are the "general position" corners, with the first two being of positive Gauss curvature and the second two being of negative Gauss curvature ("saddles"). There are then four kinds of "special position saddles," the latter

three of which are "monkey saddles." Finally, it is sometimes possible to flatten a part of one or more of the regions around an edge in the corner, as illustrated in the last two parts of figure 3.

To prove that these are the only possible types of minimizing surfaces (and to make precise what is allowed in these figures), it is useful to look at the generalized Gauss map. The Gauss map of the surface of W is essentially the dual of W: there is a vertex on S^2 corresponding to each face of W, namely the oriented unit normal to that face. Between vertices corresponding to adjacent faces of W there is a great circle segment, and each corner of W corresponds to a region of S^2 bounded by these great circle segments; in a natural way, this is the Gauss curvature of that corner. Next, one observes that if a surface is F-minimizing and consists of planar wedges containing a common vertex, and if the planar wedges are parallel to faces of W, then that surface defines an oriented cycle on the edges and vertices of the dual of W, with the edges each labelled "regular" or "inverse." Conversely, it is possible to assign a surface to every oriented, labelled cycle on the dual of W. Therefore the question of what are the F-minimizing corners can be rephrased into a question of which such cycles correspond to minimzing, embedded surfaces. For example, types (a) and (b) correspond to cycles around one face of the dual of W, with all edges labelled the same and with positive orientation, whereas types (c) and (d) correspond to any spherical triangle, with not all edges labelled the same and with negative orientation. See [5] for further details.

2C. INTERMEDIATE SURFACE TENSION FUNCTIONS.

Very little is known about the structure of F-minimizing surfaces when F is neither elliptic nor crystalline. In joint work with Cahn, however, several conjectures have been made and some experimental evidence has been found to back them up. In particular, we conjecture that a surface of least surface energy may contain a cusp, and thus that the presence of a cusp on a surface need not imply either that there is a dislocation in the body of the crystal, nor that the surface is not at equilibrium [6].

The cusp can be pictured most easily in the case that W_F is a cylinder, with its axis vertical. The cusp could then be described as ledge that peters out half-way up a vertical wall. A photograph of the surface of a sample of iron containing silicon which has a great many such ledges and several such cusps in it is shown in figure 4. This surface arose from an initially smooth surface, in conditions where the surface tension was essentially isotropic; the atmosphere was then changed, thereby altering the surface tension function. The surface is clearly not at a global equilibrium, but there is probably some sort of local almost equilibrium (see section 3A below).

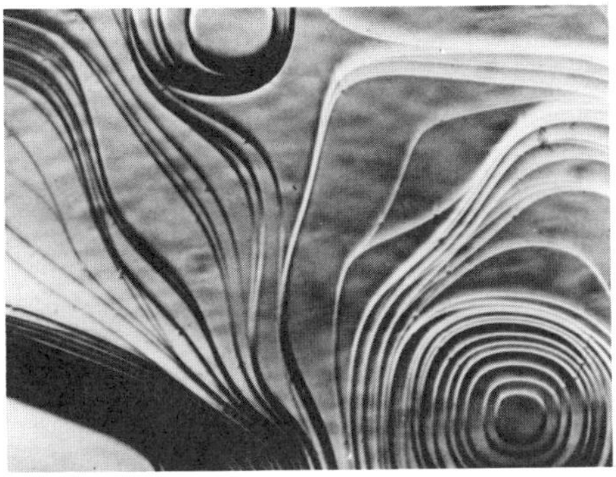

Photograph courtesy of John L. Walter,
G.E. Research and Development Center

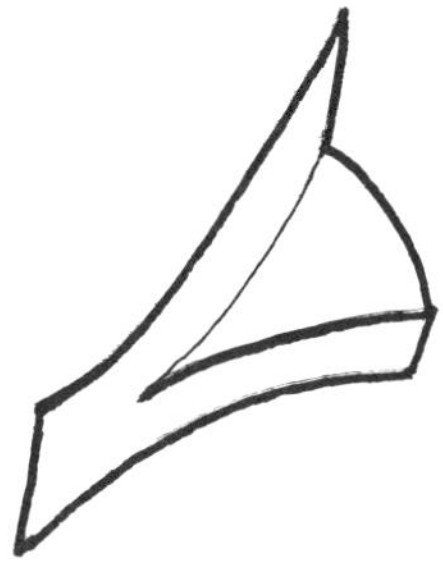

Figure 4.(a) Photograph (~2000X) of surface containing cusps; (b) cusp.

For other surface tension functions F, it is conjectured that cusps can arise corresponding to any position along a curved edge of W_F (where the tangent cone to W_F consists of two half-planes) provided W_F is flat to one side of that edge. Furthermore, it appears that when W_F is flat on neither side of a curved edge, any attempt to make a cusp will result in the surface having "infinitesimal corrugations" (being a varifold). Again, this situation is poorly understood.

3. RELATED PROBLEMS.

3A. ADDITIONAL VOLUME CONSTRAINTS AND ALMOST MINIMIZATION.

For the case of isotropic surface tension ($F \equiv$ a constant), the prescribed boundary problem has been treated in a variety of contexts, depending on the definitions of surface, boundary, and spanning a boundary. Some of the kinds of phenomena that can occur in soap films and compound soap bubbles are illustrated in [3]. The framework devised for handling the problem of area minimization with volume constraints, that of (M,ϵ,δ) minimal sets (or more generally (F,ϵ,δ) minimal sets), actually captures the idea of surfaces being locally almost minimizing and thus applies to surfaces which are not precisely at equilibrium. The definition (originally given in [2]) is as follows: Suppose B is a closed set in $\mathbf{R}^3$, $\delta > 0$, and $\epsilon: \mathbf{R}^+ \to \mathbf{R}^+ \cup \{0\}$ is nondecreasing with $\lim_{r\to 0} \epsilon(r) = 0$ (for example, if it is of the form $\epsilon(r) = Cr^\alpha$ for some $C > 0$ and $\alpha > 0$ and all $r > 0$). A set S, with $S = \operatorname{spt}(\mathcal{H}^2 \llcorner S) \sim B$ (this is a technical condition; see [2]) is **(M,ϵ,δ) minimal with respect to B** if and only if

$$\mathcal{H}^2(S \cap W) \leq (1+\epsilon(r))\, \mathcal{H}^2(\phi(S \cap W))$$

whenever ϕ: $\mathbf{R}^3 \to \mathbf{R}^3$ is Lipschitz, $W = \{x: \phi(x) \neq x\}$, $\operatorname{dist}(W \cup \phi(W), B) > 0$, and $r = \mathrm{diam}(W \cup \phi(W)) < \delta$. (The definition for (F,ϵ,δ) minimal is similar, but with F replacing H^2.) It should be emphasized that this is only a local kind of almost minimization; in particular, every C^2 manifold is (M,ϵ,δ) minimal. In [2] it was shown that (i) mathematical models for soap films and compound soap bubbles are (M,ϵ,δ) minimals sets, and (ii) any (M,ϵ,δ) minimal set is a smooth submanifold except possibly on a compact singular set of zero area. In [9] it was shown that the singular set, if present, is a finite number of smooth arcs along which three sheets of surface meet smoothly at 120 degree angles, together possibly with a finite number of vertices at which four of these arcs come together at equal (~109 degree) angles bringing together six sheets of surface.

If there are several different phases present, with all interfacial surface tensions constant but having different constant values for different interfaces, then each interface is still (M,ϵ,δ) minimal and the whole ensemble is still smooth except for a singular set of zero area [2, theorem VI.2], but one would expect that the conclusions with regard to the singular sets would be different. In particular, the 120 degree angle condition might well be modified, and more than three interfaces might meet along a curve [8]. This is a question which has not yet been satisfactorily addressed.

3B. MINIMIZATION WITH EXTERNAL FIELDS.

Wulff's construction works equally well for a crystal on a table; one simply changes the value of F in the direction pointing at the table. However, very little is known in the anisotropic case about the structure of solutions when gravity also is introduced. In fact, it is not even known if solutions must be convex bodies! If the solutions are assumed to be convex, then it can be shown that even if F is crystalline, for high enough gravity the solution may be curved near the top if F is not sufficiently symmetric [4,12].

4. CONCLUSION.

Given a problem as old as Plateau's problem, remarkably little is known about shapes when the surface tension function is assumed to be anisotropic instead of isotropic. Perhaps the reason is that soap films and bubbles have been visible to mathematicians, but the more complex versions of the analogous surfaces for crystalline materials have until relatively recently been unseen. In any case, the problem is very relevant to the structure of surfaces of materials and to the structure of grain boundaries. Furthermore, quite basic questions concerning the way surfaces change when the surface tension function changes (for example, through changing the temperature or surrounding atmosphere) remain unanswered. Finally, the existence and even necessity of varifold solutions (generalized surfaces) indicates that edge energies really must be introduced into the model.

REFERENCES

[1] W. K. Allard, *On the first variation of a varifold*, Ann. of Math. **95** (1972), 417-491.

[2] F. J. Almgren, Jr., *Existence and regularity almost everywhere of solutions to elliptic variational problems with constraints*, Mem. A. M. S. **4**, Number 165.

[3] F. J. Almgren and J. E. Taylor, *The geometry of soap bubbles and soap films*, Scientific American **235** (July 1976), 82-93.

[4] J. Avron, J. E. Taylor, and R. K. Zia, *Equilibrium shapes of crystals in a gravitational field*, J. Stat. Phys. **33** (1983), to appear.

[5] J. W. Cahn and J. E. Taylor, *A catalog of saddle shaped surfaces in crystals*, in preparation.

[6] J. W. Cahn and J. E. Taylor, *A cusp singularity in surfaces*, in preparation.

[7] A. Dinghas, *Uber einen geometrischen Satz von Wulff fur die Gleichgewichtsform von Kristallen*, Zeitschrift fur Kristallographie **105** (1944), 304-314.

[8] D. W. Hoffman and J. W. Cahn, *A vector thermodynamics for anisotropic surfaces - I. Fundamentals and applications to plane surface junctions*, Surface Sci. **31** (1972), 368-388.

[9] J. E. Taylor, *The structure of singularities in soap-bubble-like and soap-film-like minimal surfaces*, Annals of Math. **103** (1976), 489-539.

[10] J. E. Taylor, *Crystalline variational problems*, Bull. Amer. Math Soc. **84** (1978), 568-588.

[11] J. E. Taylor, *Constructing crystalline minimal surfaces*, Annals of Math. Studies **105**, Seminar on Minimal Submanifolds, E. Bombieri, Ed. (1983), 271-288.

[12] J. E. Taylor, *Is there gravity induced facetting of crystals?*, Proceedings of C.I.R.M. Congress on "Variational methods for equilibrium problems of fluids," to appear.

[13] G. Wulff, *Zur Frage der Geschwindigkeit des Wachsthums und der Auflosung der Krystallflachen*, Zeitschrift fur Krystallographie und Mineralogie **34** (1901), 499-530.

[14] L. C. Young, *Generalized surfaces in the calculus of variations I, II*, Ann. of Math. **43** (1942), 84-103,530-544.

The author was partially supported by National Science Foundation Grant MCS-8301869.

Mathematics Department
Rutgers University
New Brunswick, NJ 08903

INDEX